PRACTICE MAKES PERMANENT

300+ questions

AQA GCSE Biology

Jo Ormisher

HODDER EDUCATION
AN HACHETTE UK COMPANY

The Publishers would like to thank the following for permission to reproduce copyright material.

Photo credits

Page 7, figure 6 © Sinhyu/stock.Adobe.com

Acknowledgements

Every effort has been made to trace all copyright holders, but if any have been inadvertently overlooked, the Publishers will be pleased to make the necessary arrangements at the first opportunity.

Although every effort has been made to ensure that website addresses are correct at time of going to press, Hodder Education cannot be held responsible for the content of any website mentioned in this book. It is sometimes possible to find a relocated web page by typing in the address of the home page for a website in the URL window of your browser.

Hachette UK's policy is to use papers that are natural, renewable and recyclable products and made from wood grown in well-managed forests and other controlled sources. The logging and manufacturing processes are expected to conform to the environmental regulations of the country of origin.

Orders: please contact Hachette UK Distribution, Hely Hutchinson Centre, Milton Road, Didcot, Oxfordshire, OX11 7HH. Telephone: +44 (0)1235 827827. Email education@hachette.co.uk Lines are open from 9 a.m. to 5 p.m., Monday to Friday. You can also order through our website: www.hoddereducation.com

ISBN: 978 1 5104 7642 4
© Jo Ormisher 2020
First published in 2020 by
Hodder Education
An Hachette UK Company,
Carmelite House, 50 Victoria Embankment
London EC4Y 0LS

Impression number 5 4 3 2

Year 2024 2023

Cover photo © Sergey Skleznev/stock.Adobe.com

Illustrations by Integra.

Cell biology figures 2, 3, 8, 10, 12; Organisation figures 2, 3, 5, 9, 11, 12, 15; Infection and response figures 1, 5; Bioenergetics figures 1, 2, 4, 5; Homeostasis and response figures 2-5, 7-9; Inheritance, variation and evolution figure 5; Ecology figure 4; Practice exam paper 1 figures 2-4; Practice exam paper 2 figure 4 © Aptara

Typeset in Integra Software Services Pvt. Ltd., Pondicherry, India.

Printed and bound by CPI Group (UK) Ltd, Croydon CR0 4YY

A catalogue record for this title is available from the British Library.

Contents

Introduction

Practice Makes Permanent is a series that advocates the benefits of answering lots and lots of questions. The more you practise, the more likely you are to remember key concepts; practice does make permanent. The aim is to provide you with a strong base of knowledge that you can automatically recall and apply when approaching more difficult ideas and contexts.

This book is designed to be a versatile resource that can be used in class, as homework, or as a revision tool. The questions may be used in assessments, as extra practice, or as part of a SLOP (i.e. Shed Loads of Practice) teaching approach.

How to use this book

This book is suitable for the AQA GCSE Biology course, both at Higher and Foundation levels. It covers all the content that you will be expected to know for the final examination.

The content is arranged topic-by-topic in the order of the AQA specification, so areas can be practised as needed. Within each topic there are:

- **Quick questions** – short questions designed to introduce the topic.
- **Exam-style questions** – questions that replicate the types, wording and structure of real exam questions, but highly-targeted to each specification point.
- **Topic reviews** – sections of exam-style questions that test content from across the entirety of each topic more synoptically.

These topic questions are tagged with the following:

p64	page references for the accompanying Hodder Education Student Book: AQA GCSE (9-1) Biology, 9781471851339. This can be revisited before or after attempting the questions in a topic.
4.1.1.1	the AQA specification reference, which can be used if you want to practise specific areas.
H	indicates Higher-only content.
MS5b	indicates where questions test Maths skills.
QWC	indicates where answers will also be marked on the quality of written communication.
WS	indicates where questions require you to work scientifically.
AT	indicates where questions ask you to use practical knowledge of apparatus and techniques.
RP	indicates where questions test understanding of required practicals.

At the end of the book there is a full set of **practice exam papers**. These have been carefully assembled to resemble typical AQA question papers in terms of coverage, marks and skills tested. We have also constructed each one to represent the typical range of demand in the GCSE Biology specification as closely as possible.

Full worked **answers** are included at the end of the book for quick reference, with awarded marks indicated where appropriate.

Cell biology

Cell structure

Quick questions

p2	4.1.1.1	**1** What is a 'eukaryotic cell'?
p3	4.1.1.1	**2** What is a 'prokaryotic cell'?
p5	4.1.1.1	**3** Are plant cells eukaryotic or prokaryotic?
p3	4.1.1.1	**4** Are bacterial cells eukaryotic or prokaryotic?
p6	4.1.1.2	**5** Plant and algal cells have cell walls. What is their cell wall made of?
p10	4.1.1.4	**6** Cells **differentiate** as an organism develops. What does differentiate mean?
p12	4.1.1.5	**7** What is meant by the 'resolution' (or 'resolving power') of a microscope?
p12	4.1.1.5	**8** Why can ribosomes not be seen using a light microscope?
p12	4.1.1.5	**9** Microscopes were first invented in 1590. Give the reason why mitochondria were not seen until 1840 and ribosomes were not seen until 1953.
p15	4.1.1.5	**10** Give the formula for calculating the magnification of an object.
p11	4.1.1.5	**11** Rearrange the equation for magnification to find:

11 Rearrange the equation for magnification to find:

- the real size of an object
- the image size.

p11	4.1.1.5 WS4.4, 4.5 MS1b	**12** How many of the following are there in 1 metre?

- millimetres (mm)
- micrometres (μm)
- nanometres (nm).

Give your answers in standard form.

p91	4.1.1.6 RP2 MS5c	**13** Give the equation used to calculate the area of a circle.
	4.1.1.6 MS1b **Ⓗ**	**14** Convert these numbers into standard form:

- 456 000
- 0.00032

Exam-style questions

15 **Figure 1** shows two cells labelled A and B. One is a prokaryotic cell and one is a eukaryotic cell.

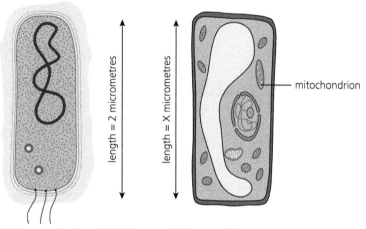

Cell A, magnification × 40 000   Cell B, magnification × 400

Figure 1

p3 **4.1.1.1**

15–1 Give the letter of the prokaryotic cell. [1]

p3–5 **4.1.1.1**

15–2 Describe **two** ways that cell **A** is different from cell **B**. [2]

p16 **4.1.1.1** **WS4.5** **MS2h**

15–3 Cell **A** is 2 micrometres (μm) long. Give its length in millimetres (mm) and in nanometres (nm). [2]

p15 **4.1.1.5** **MS2h, 3b** **WS4.5**

15–4 **Figure 1** shows cell **A** and cell **B** the same length, but the magnification of each cell is different.

Cell **A** is 2 μm long. Cell **B** is X μm long.

Calculate the length of cell **B**. [2]

p4–5 **4.1.1.2**

15–5 Mitochondrion, ribosome and nucleus are structures found in eukaryotic cells.

Write the structures in order of size from smallest to largest. [1]

p3–5 **4.1.1.1**

15–6 Suggest **one** reason why prokaryotic cells do not have mitochondria.

Use information from **Figure 1**. [1]

Total: 9

16 Cells are the basic unit of all living things.

Cells are either eukaryotic or prokaryotic.

Eukaryotic and prokaryotic cells have different structures.

p3–6 **4.1.1.1**

16–1 Copy and complete **Table 1** to show the structures present in each type of cell. [3]

Put a tick or a cross in each box.

	Prokaryotic cells only	Eukaryotic cells only	Prokaryotic and eukaryotic cells
Cell membrane			
Cell wall			
Cytoplasm			
Nucleus			
Plasmid			

Table 1

p3–6 | 4.1.1.1

16–2 Eukaryotic cells and prokaryotic cells both contain genetic material.

Describe **two** ways that the genetic material is arranged differently in prokaryotic and eukaryotic cells. [2]

p3–6 | 4.1.1.2

16–3 Cells contain sub-cellular structures.

Match the names of the sub-cellular structures, 1–5, with their function, A–E. [5]

Name of sub-cellular structure		Function
1 Chloroplasts		A Site of aerobic respiration
2 Mitochondria		B Gives structure and support to the cell
3 Cell wall		C Site of protein synthesis
4 Permanent vacuole		D Site of photosynthesis
5 Ribosome		E Supports the cell, filled with cell sap

p5–6 | 4.1.1.2

16–4 Plant cells contain chloroplasts, but animal cells do not.

Give **two** other differences between plant cells and animal cells. [2]

p12 | 4.1.1.5

16–5 Some of the sub-cellular structures in cells can only be seen using an electron microscope.

Give **two** reasons why electron microscopes are used to study cells in finer detail than light microscopes. [2]

p10 | 4.1.1.4

16–6 Cells may differentiate to become specialised cells.

Explain what happens when a cell differentiates. [2]

p10 | 4.1.1.4

16–7 Describe the main difference between differentiation in animal cells and plant cells. [2]

p8–10 &69 | 4.1.1.3 | QWC

16–8 Describe at least **three** of the structures and functions of specialised plant and animals cells. [6]

Total: 24

p4–7&11 | 4.1.1.2

17 **Figure 2** shows a light microscope.

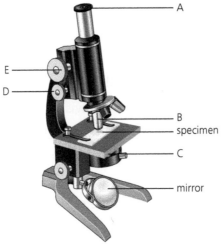

Figure 2

17–1 Name the parts of the microscope labelled A–E. Choose your answers from the options below. [5]

eyepiece lens	objective lenses	coarse focus	fine focus	stage

RP1
AT7

17–2 Describe a method used to prepare cheek cells for viewing with a light microscope. [3]

QWC
RP1
AT7

17–3 Describe how a light microscope can be used to view a prepared slide of cells at **high** power. [6]

WS1.2, 3.1
AT7

17–4 **Figure 3** shows a cheek cell seen with a light microscope.

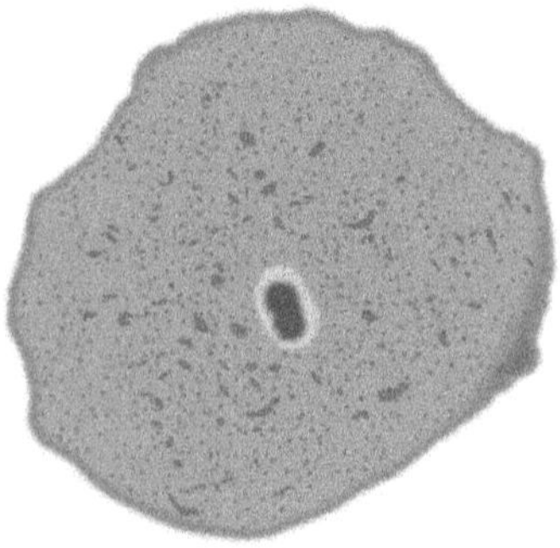

Figure 3

Draw the cell shown in the photo. Label the cell membrane, cytoplasm and nucleus. [2]

17–5 Describe the functions of these three parts of a cell: cell membrane, cytoplasm and nucleus. [3]

17–6 The cell in **Figure 3** is an animal cell.

Give **two** reasons why the cell can be identified as an animal cell and not a plant cell. [2]

RP1
AT7

17–7 A student uses a light microscope to look at cheek cells. The image is not clear.

Suggest what the student needs to do to the microscope to produce a clear image. [1]

RP1
AT7

17–8 Another student looks at cheek cells using the light microscope, but cannot see individual cells.

Suggest what the student needs to do to the microscope to see individual cells. [2]

Total: 24

p80–2 4.1.1.6

18 Bacteria can divide very rapidly.

18–1 Name the process of simple cell division in bacteria. [1]

18–2 Give **two** conditions needed for rapid cell division. [2]

18–3 Bacteria can be grown in a culture medium. Name **one** example of a culture medium. [1]

WS2.2, 2.4
RP2
AT4
QWC

18–4 A student is given a pure culture of bacteria in liquid nutrient broth.

Describe how the student can prepare an uncontaminated culture of bacteria on solid agar jelly using aseptic technique. Explain why each of the steps is necessary. [6]

WS3.3, 4.6
MS1b, d,2a

18–5 The student uses the uncontaminated culture to investigate the effect of antibiotics.

The student grows the bacteria on an agar plate with paper discs containing antibiotics, then measures the diameters of the zones of inhibition around each paper disc.

The bacteria the student uses has a mean division time of 40 minutes. Starting with a single bacterial cell, calculate the number of bacteria in a population after 24 hours.

Give your answer in standard form. *[3]*

MS1c

18–6 **Table 2** shows the diameters of the zones of inhibition for four antibiotics tested by the student.

Antibiotic	Diameter of zone of inhibition in mm	Area of zone of inhibition in mm²
A	12	
B	8	50
C	0	0
D	6	28

Table 2

Calculate the area of the zone of inhibition for antibiotic A. *[2]*

WS3.5

18–7 Which antibiotic is the most effective? Give **one** reason for your choice. *[2]*

Total: 17

Cell division

Quick questions

p19 | 4.1.2.1

1 Name the part of the cell that contains chromosomes.

p19 | 4.1.2.1

2 What are chromosomes made of?

p19 | 4.1.2.1

3 What are carried on chromosomes?

p19 | 4.1.2.1

4 How many of each chromosome is found in a human body cell?

Exam-style questions

p19–23 | 4.1.2.1

5 New cells are produced by cell division.

Figure 4 shows an animal cell with some of its structures magnified to show more detail.

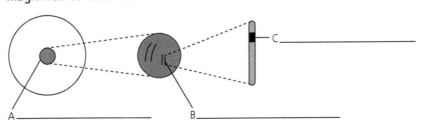

Figure 4

5–1 Name parts A–C. Choose your answers from the options below. *[3]*

chromosome	gene	nucleus

5–2 Multicellular organisms, such as plants, use cell division during their development.

Give **one** other use of cell division by mitosis in multicellular organisms. *[1]*

5–3 Plants contain meristem tissue. What is the function of meristem tissue? *[1]*

5–4 Stem cells from meristem tissue in plants can be used to produce clones.

Give **two** advantages of producing clones using stem cells from meristem tissue. *[2]*

5–5 Plant cloning can be used to protect rare species from extinction. Describe one other use of plant cloning. *[2]*

Total: 9

p20–1 4.1.2.2

6 Cells divide in a series of stages called the cell cycle.

Before a cell can divide, changes need to happen in the cell. One of the changes is to the genetic material.

Figure 5 shows a chromosome before and after one of the stages of the cell cycle.

Figure 5

6–1 What process has caused the change in the appearance of the chromosome? *[1]*

6–2 Give **one** other change that happens in a cell before it divides that is not related to its genetic material. *[1]*

6–3 A cell from an onion has eight chromosomes. The cell divides by mitosis.

Give the number of chromosomes in one of the new cells. *[1]*

6-4 **Figure 6** shows some onion cells at different stages of the cell cycle.

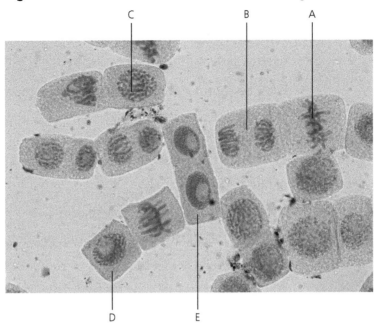

Figure 6

Give the letter of the cell that is **not** dividing by mitosis. *[1]*

6-5 What is happening to the chromosomes in cell B? *[1]*

6-6 Describe what is happening in cell E. *[2]*

Total: 7

p22-4 4.1.2.3

7 Stem cells have an important function in living organisms.

7-1 What is a 'stem cell'? *[2]*

7-2 Human stem cells can be found in human embryos and in adult bone marrow.

Describe the main difference between stem cells from embryos and stem cells from adult bone marrow. *[2]*

7-3 Therapeutic cloning produces an embryo with the same genes as the patient.

Give **one** advantage of treatment with cells that have the same genes as the patient. *[1]*

7-4 Name **two** conditions that could be treated with stem cells. *[2]*

WS3 7-5 Describe **two** reasons why people may be against the use of stem cells. *[2]*

7-6 Leukaemia is a disease that affects the blood. A patient with leukaemia can be treated using stem cells. The stem cells can be obtained from the patient's own bone marrow. Stem cells can also be obtained from human embryos.

WS1.3
QWC
Evaluate the use of stem cells from the patient and from human embryos. *[6]*

Total: 15

Transport in cells

Quick questions

1 Write a definition of **diffusion**.

2 What do substances move across to get into and out of cells?

3 Name **two** substances that move into cells and **two** substances that move **out of** cells.

4 Name **two** plant organs that are adapted for exchanging materials.

5 Define the term osmosis.

6 What does 'partially permeable' mean?

7 Write out the equation for calculating percentage change in mass.

8 Define the term active transport.

9 Plants require mineral ions for healthy growth. Where do plants get mineral ions from?

10 Which process do plants use to take up mineral ions from very dilute solutions?

11 Which molecules are absorbed into the blood by active transport and used for cell respiration?

Exam-style questions

12 An organism's surface area to volume ratio affects its ability to transport sufficient molecules into and out of its cells.

Figure 7 shows cubes with different side lengths.

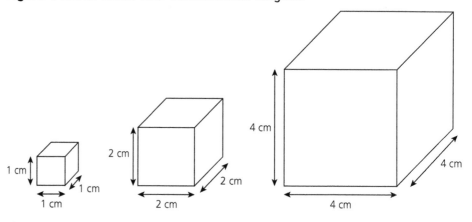

Figure 7

Surface area to volume ratio can be calculated using the equation:

$$\text{surface area to volume ratio} = \frac{\text{surface area}}{\text{volume}}$$

MS1c,5c
12–1 Explain how surface area to volume ratio changes as the length of the sides double.

Use calculations to support your answer. [3]

WS3.5
12–2 Compare the surface area to volume ratio of a single-celled organism with the surface area to volume ratio of a large multicellular organism. [1]

12–3 Explain how single-celled organisms obtain the oxygen they need for aerobic respiration. [2]

12–4 Insects do not have lungs or blood. Suggest how insects obtain the oxygen they need for aerobic respiration. [2]

12–5 Explain how large multicellular organisms, such as mammals, obtain the oxygen they need to carry out aerobic respiration. [2]

Total: 10

p31-2& 44   4.1.3.1

13 Substances may move into and out of the cells of living organisms by diffusion.

The rate of diffusion can be affected by the concentration gradient.

13–1 Name **two** other factors that affect the rate of diffusion. [2]

13–2 Explain how a change in concentration gradient affects the rate of diffusion. [3]

13–3 The lungs of a mammal are adapted for gas exchange.

Describe **two** ways that a concentration gradient is maintained in the lungs. [2]

13–4 The small intestine of mammals is adapted to absorb digested food. The small intestine contains finger-like projections called villi. **Figure 8** shows one villus.

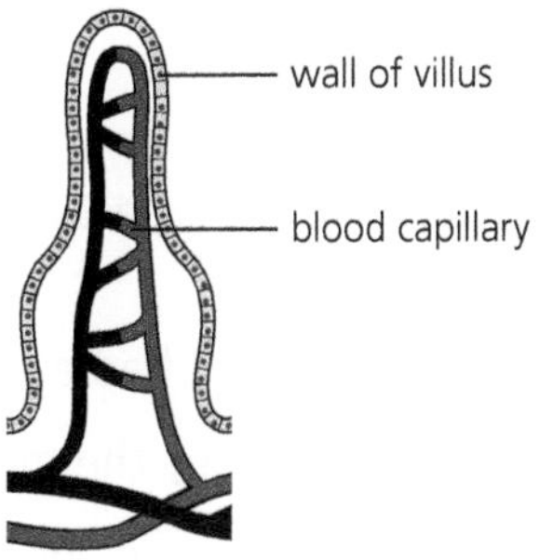

wall of villus

blood capillary

Figure 8

Explain **one** way that the small intestine is adapted for its function. [2]

13–5 Fish have gills for gas exchange. The gills are made of many gill filaments. The outer layer of each gill filament is one cell thick. Each gill filament contains a blood capillary.

Explain **two** adaptations of the gills for rapid gas exchange. [2]

Total: 11

p33–4 4.1.3.2

14 A student investigates osmosis using 'bags' made from an artificial partially permeable membrane, and different concentrations of sugar solution.

The student:

- fills three 'bags' with 10 cm³ of 5% sugar solution
- finds the mass of each 'bag'
- puts each bag in a boiling tube with a different concentration of sugar solution in each tube
- waits for 30 minutes
- finds the mass of each bag again.

Figure 9 shows how the student sets up the apparatus.

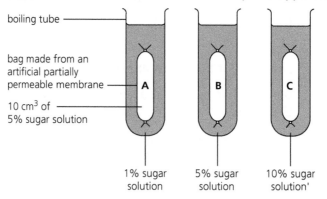

Figure 9

WS1.2

14–1 Give the reason why the student used a partially permeable membrane in the investigation. *[1]*

MS1c

14–2 At the start of the investigation, bag **C** had a mass of 10.7 g.

After 30 minutes in 10% sugar solution, the mass of bag **C** was 9.8 g.

Calculate the percentage change in mass of bag **C**. *[2]*

WS1.2

14–3 Explain why bag **C** decreased in mass. *[2]*

WS1.2

14–4 Describe and explain what will happen to the masses of bags **A** and **B**. *[4]*

Total: 9

p35 4.1.3.2 RP3

15 A student investigates osmosis by measuring the effect of different concentrations of salt solution on the mass of carrot cylinders.

This is the method the student uses:

1 Label six boiling tubes 0.0 M, 0.2 M, 0.4 M, 0.6 M, 0.8 M and 1.0 M.

2 Add 10 cm³ of each concentration of salt solution to the boiling tubes.

3 Use a cork borer to cut six carrot cylinders, then trim the cylinders to the same length.

4 Gently dry the carrot cylinders with a paper towel.

5 Measure the mass of each carrot cylinder.

6 Put one carrot cylinder into each boiling tube for one hour.

7 Remove the carrot cylinders from the test tubes and gently dry them with a paper towel.

8 Measure the mass of each carrot cylinder.

WS2.2 **15–1** What is the independent variable for this investigation? [1]

WS2.2 **15–2** What is the dependent variable for this investigation? [1]

WS2.2 **15–3** Give **two** variables that the student controlled in this investigation. [2]

WS2.3 **15–4** Give the reason why the student dries the carrot cylinders. [1]

WS3.1
MS4a,4c **15–5** Describe how the student could use a graph to find the concentration of salt inside the carrot cells. [3]

Total: 8

16 Substances move into and out of cells by active transport, diffusion or osmosis.

p29–36 4.1.3.3 **16–1** Describe **two** similarities between diffusion and osmosis. [2]

p29–36 4.1.3.3 **16–2** Describe **two** differences between active transport and osmosis. [2]

p9 4.1.1.3, 4.1.3.1 **16–3** Root hair cells absorb water and mineral ions from the soil.

Figure 10 shows a root hair cell.

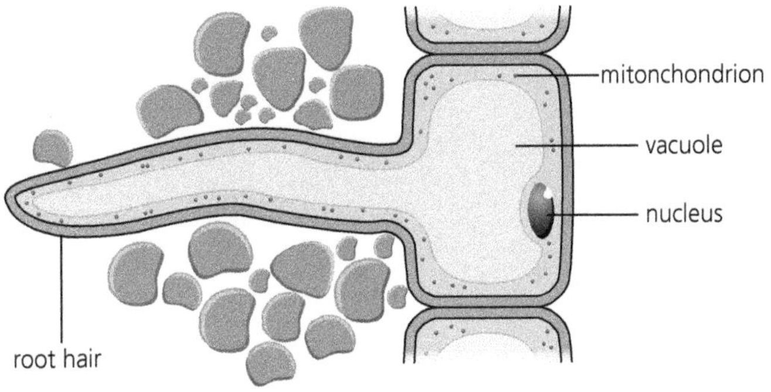

Figure 10

Describe how the shape of the root hair cell increases the rate of water uptake. [2]

p36 4.1.3.3 **16–4** Mineral ions are in a more dilute solution in the soil than inside the cell.

Name the process that transports mineral ions into the root hair cell. [1]

p5&36 4.1.3.3 **16–5** The root hair cell has large numbers of mitochondria.

Explain how this increases the uptake of mineral ions. [3]

p36 4.1.3.3 **16–6** Some soils are water-logged. This means that there is water filling the air spaces in the soil.

Explain why plants grown in water-logged soil are deficient in mineral ions. [3]

Total: 13

Cell biology topic review

1   A student investigates the effect of a range of concentrations of salt solution on the mass of potato cylinders.

The student follows this method:

- Cut six potato cylinders to the same length and diameter.
- Carefully blot the cylinders dry with a paper towel.
- Weigh each cylinder.
- Put one cylinder into each boiling tube (**Figure 11**).
- Remove the cylinders from the tubes after one hour.
- Carefully dry each cylinder and reweigh them.

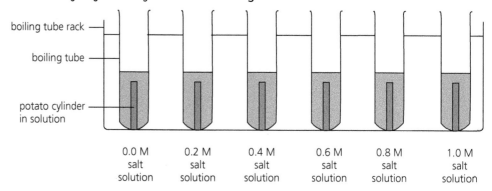

boiling tube rack

boiling tube

potato cylinder in solution

0.0 M salt solution   0.2 M salt solution   0.4 M salt solution   0.6 M salt solution   0.8 M salt solution   1.0 M salt solution

Figure 11

Table 3 shows the student's results.

Concentration of salt solution in M	Mass of potato at start in g	Mass of potato at end in g	Change in mass in g	Percentage change in mass in %
0.0	19.6	20.9	+1.3	+6.6
0.2	16.8	17.2	+0.4	+2.4
0.4	22.1	20.5	−0.6	−2.7
0.6	17.0	14.7	−2.3	
0.8	19.2	16.4	−2.8	−14.6
1.0	27.9	23.6	−4.3	−15.4

Table 3

WS1.2, 3.5
RP2

1–1   Explain why there was an increase in mass for the salt concentrations 0.0 M and 0.2 M.   *[2]*

MS1c

1–2   Calculate the percentage change in mass for the salt concentration 0.6 M.   *[1]*

WS1.2

1–3   Give **one** reason why the student calculated the percentage change in mass.   *[1]*

MS4c

1–4   Plot a graph of the student's results.   *[4]*

- Choose suitable scales.
- Label the axes.
- Plot the percentage change in mass for each concentration of salt solution.
- Draw a line of best fit.

MS4a   1–5   Use your graph to find the concentration of salt solution where there is no change in mass.   [1]

WS1.2   1–6   What is the significance of the concentration of salt solution that gives no change in mass?   [1]

Total: 10

2   **Figure 12** shows a plant cell.

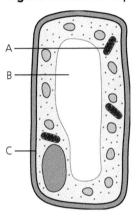

Figure 12

p5   4.1.1.2   2–1   Name the structures labelled A, B and C.   [3]

p5   4.1.1.2   2–2   Give the function of ribosomes.   [1]

p4   4.1.1.1   2–3   Give the evidence from **Figure 12** that the plant cell is eukaryotic.   [1]

p15–16   4.1.1.5   WS4.4, 4.5   2–4   The real length of the plant cell is 90 micrometres (µm). Calculate the magnification of the plant cell in **Figure 12**.   [2]
MS3d

p6   4.1.1.2   RP1   2–5   A student looks at some plant cells using a light microscope. Describe how the student can use the light microscope to produce a clear image of the cells at a higher magnification.   [2]
WS2.3

Total: 9

3   **Figure 13** shows a cell from the lining of the small intestine.

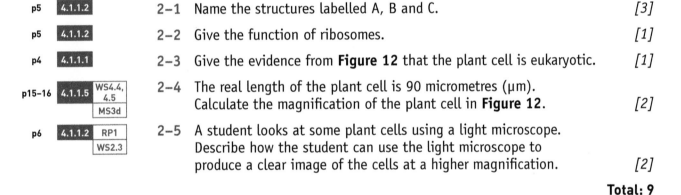

cell membrane folded to form microvilli

mitochondria

Figure 13

The cell is specialised to absorb digested food; for example, glucose (sugar).

p31 4.1.1.2, 4.1.1.3, 4.1.3.3

3–1 Explain why the cell has lots of mitochondria. [3]

p31 4.1.3.1

3–2 The cell membrane is folded to form microvilli. Explain why this would increase the rate of absorption. [2]

p20&22 4.1.1.4, 4.1.2.2, 4.1.2.3

3–3 Explain how stem cells are able to produce all of the cells required in a growing foetus. [3]

Total: 8

4 **Figure 14** shows an axolotl.

gills

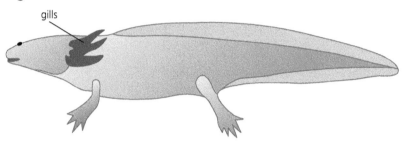

Figure 14

Axolotls are multicellular animals that live in water.

Axolotls have gills for gas exchange.

p31&33 4.1.3.1

4–1 Explain why axolotls use gills rather than conducting gas exchange through their epidermis. [2]

p31&33 4.1.3.1 WS3.5

4–2 Suggest **two** features of an axolotl's gills that make it an effective gas exchange surface. [2]

p22 4.1.1.4, 4.1.2.3

4–3 If an axolotl's leg is removed, it can regrow a new leg.

To regrow a new leg, the specialised cells in the axolotl have to dedifferentiate to produce cells that can form new bone, muscle or skin cells.

Name the type of cells produced when specialised cells dedifferentiate. [1]

p20–1 4.1.2.2 WS4.2, 4.4

4–4 The new bone, muscle and skin cells divide by mitosis to grow a new leg. DNA replicates during mitosis.

The mass of DNA in an axolotl skin cell is 36 picograms (1 picogram = 10^{-12} grams).

- What is the mass of DNA in the skin cell after the DNA replicates? [1]
- What is the mass of DNA in one of the new skin cells? [1]

p23 4.1.2.3

4–5 Scientists use axolotls in research to find ways of regenerating human tissues and organs.

Name **one** condition that this research could find a treatment for. [1]

p24 4.1.2.3 WS1.3

4–6 Suggest **one** reason why some people may be against this type of research. [1]

Total: 9

Organisation

Principles of organisation

Quick questions

p42 | 4.2.1

1 What are the basic building blocks of all living organisms?

p42 | 4.2.1

2 What is a 'tissue'?

p42 | 4.2.1

3 What is an 'organ'?

Exam-style questions

p42 | 4.2.1

4–1 Arrange these structures in size order, from smallest to largest.   [2]

| Cell | Organ | Organ system | Organism | Tissue |

p42& 68–9 | 4.2.1

4–2 Give **two** examples of tissues.   [2]

p42& 70–1 | 4.2.1

4–3 Give **two** examples of organs.   [2]

p42&71 | 4.2.1

4–4 Organs are organised to form organ systems. Give **two** examples of organ systems.   [2]

Total: 8

Animal tissues, organs and organ systems

Quick questions

p42 | 4.2.2.1

1 Describe the function of the digestive system.

p42&45 | 4.2.2.1

2 Define digestion and name the **three** main types of digestive enzymes.

p48 | 4.2.2.1

3 What are 'enzymes' and how do they work?

p51 | 4.2.2.2

4 Describe the function of the heart.

p52 | 4.2.2.2

5 Explain what 'double circulation' means.

p56–7 | 4.2.2.2

6 Where are coronary arteries found and what is their function?

p51&57 | 4.2.2.2

7 How can a problem with a person's pacemaker be treated?

p54–6 | 4.2.2.2

8 Name the components of blood tissue and give their functions.

p57 | 4.2.2.4

9 Explain how stents are used to treat coronary heart disease (CHD).

p56–7 | 4.2.2.4

10 Describe how an unhealthy lifestyle affects the coronary arteries and list ways of reducing the risk of CHD.

p58–9 | 4.2.2.5

11 Explain what is meant by the term 'health' and list ways of improving health.

p61–2 `4.2.2.6`

12 Explain what is meant by the term 'risk factor', and give examples of risk factors and their associated diseases.

pXX `4.2.2.6`

13 What is meant by 'obesity'?

p62 `4.2.2.6`

14 Define the term 'carcinogen' and give **two** examples.

p60 `4.2.2.7`

15 Describe how cell division can result in cancer.

Exam-style questions

16 The digestive system digests and absorbs food.

Figure 1 shows the human digestive system.

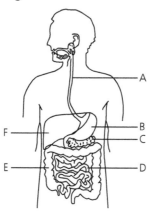

Figure 1

p43 `4.2.2.1`

16–1 Name the parts of the digestive system labelled A–F. [6]

p43–5 `4.2.2.1`

16–2 Match the part of the digestive system, 1–4, to its function, A–D. [4]

Part of digestive system	Functions
1 Oesophagus	A Absorption of soluble products of digestion.
2 Stomach	B Absorption of water and mineral ions.
3 Small intestine	C Transport of food to stomach by peristalsis.
4 Large intestine	D Production of hydrochloric acid.

p47 `4.2.2.1` AT8 QWC WS2.2, 2.3

16–3 The digestive system digests carbohydrates, lipids and proteins.

Describe how a student could test a sample of food to find out whether it contains carbohydrates, lipids and proteins. [6]

p49 `4.2.2.1`

16–4 Digestion is completed by digestive enzymes.

What is the main function of all digestive enzymes? [1]

p43–6 `4.2.2.1`

16–5 Copy and complete boxes **A–E** in **Table 1** to show the sites of production and the effects of the digestive enzymes. [5]

Digestive enzyme	Site(s) of production	Substrate	Product(s)
Amylase (carbohydrase)	Salivary glands and pancreas	A	Sugars
Lipase	B	Lipids (fats and oils)	C
Protease	Stomach, pancreas and small intestine	D	E

Table 1

p46 `4.2.2.1`

16–6 The liver produces bile that is stored in the gall bladder. Describe the functions of bile. [4]

Total: 26

p48–9 4.2.2.1

17 **Figure 2** shows the 'lock and key' model of enzyme action.

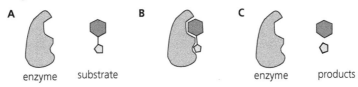

Figure 2

WS1.2 | **17–1** Describe what is happening in parts **A** to **C** in **Figure 2**. [4]

QWC | **17–2** Enzyme activity can be affected by factors including temperature and pH.

Figure 3 shows the effect of temperature on enzyme activity.

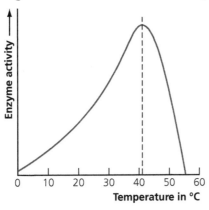

Figure 3

QWC | **17–2** Describe and explain how increasing the temperature affects the rate of reaction. [6]

RP5
WS2.2
AT1,2,5,8 | **17–3** A student investigates the effect of pH on the rate of reaction of the enzyme amylase.

This is the method the student uses:

- Add 2 cm³ of pH 4 buffer solution and 4 cm³ of starch solution to a test tube and put it in a water bath at 35 °C for 2 minutes.

- Add 2 cm³ of amylase solution to another test tube and put it in the water bath for 2 minutes.

- Put one drop of iodine solution into each well on a spotting tile.

- Pour the amylase solution into the test tube with starch and buffer solutions. Start a stopwatch and stir the mixture with a glass rod.

- Remove one drop of the mixture with a glass rod every 30 seconds, and put it in a well of the spotting tile with the iodine solution.

- Keep sampling every 30 seconds until the iodine solution does not change colour.

- Repeat the investigation with buffer solutions at pH 5–9.

What is the dependent variable for this investigation? [1]

RP5
WS2.2
AT1,2,5,8

17–4 The student controlled the temperature in this investigation.

Give **one** other variable that the student controlled. [1]

RP5
WS2.2
AT1,2,5,8

17–5 How did the student control the temperature? [1]

RP5
WS2.2
AT1,2,5,8

17–6 Explain why temperature must be controlled in the investigation. [2]

RP5
WS2.2
AT3

17–7 Suggest why measuring the concentration of starch every 30 seconds would be better than recording a colour change. [1]

RP5
WS2.2
AT1

17–8 Suggest **one** way the student could more accurately measure the time taken for all of the starch to break down. [1]

17–9 The student repeats the investigation using a pH 2 buffer solution, starch solution and amylase solution. After 15 minutes, the iodine solution still turns blue–black. Explain why. [2]

Total: 19

18   **Figure 4** shows the human breathing system.

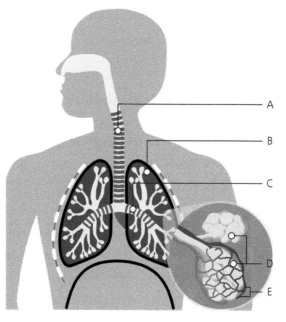

Figure 4

p31   4.2.2.2

18–1 Name the parts of the breathing system labelled A–E. Choose your answers from the options below. [5]

Alveoli	Bronchus	Capillary network	Lung	Trachea

p61–2   4.2.2.6

18–2 Smoking tobacco is harmful. Smoking can increase the risk of lung diseases including lung cancer.

Explain why smoking increases the risk of lung cancer. [3]

p62   4.2.2.6

18–3 Name **two** lung diseases other than cancer that smoking increases the risk of. [2]

p56 4.2.2.6   18–4 Smoking can affect the cardiovascular system. Give **two** ways that smoking can increase the risk of cardiovascular disease. [2]

p62 4.2.2.6   18–5 Women who smoke when they are pregnant can affect their unborn babies. Describe **two** possible effects of smoking on unborn babies. [2]

Total: 14

19 **Figure 5** shows a cross-section through a human heart.

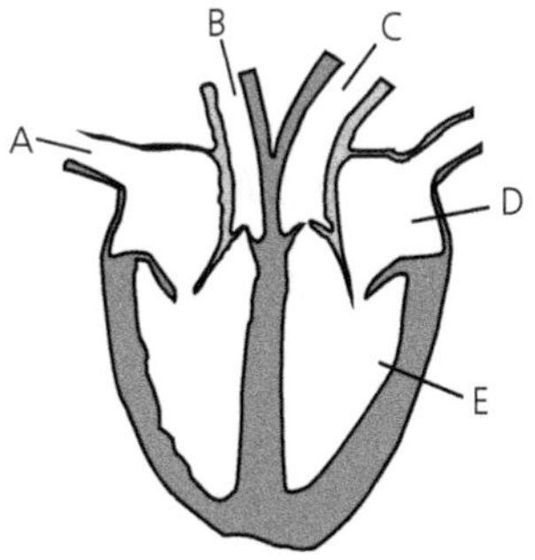

Figure 5

p51 4.2.2.2   19–1 Name the parts of the heart labelled A–E. Choose your answers from the options below. [5]

Left atrium	Left ventricle	Right atrium	Right ventricle
Aorta	Pulmonary artery	Pulmonary vein	Vena cava

p51 4.2.2.2   19–2 State the location and the function of the 'pacemaker' of the heart. [2]

p53 4.2.2.2   19–3 Give the functions of the left and right ventricles. [2]

p56&62 4.2.2.6   19–4 The coronary arteries supply blood to the heart muscle.

The coronary arteries can be affected by coronary heart disease (CHD).

Name **three** factors associated with an increased risk of CHD. [3]

p56 4.2.2.4   19–5 Explain what happens to the heart in CHD. [3]

Total: 15

p57 4.2.2.4   20 The heart contains valves to prevent the backflow of blood.

20–1 Give **one** heart valve fault that can develop in some people. [1]

20–2 Describe the effect a faulty valve would have on a person. [2]

20–3 How can faulty valves be treated? [1]

20–4 Sometimes the heart is unable to pump blood around the body properly. This is called heart failure.

Heart failure can be treated using an artificial heart, or by organ transplant.

Give **one** other situation when an artificial heart might be used. [1]

WS1.3, 1.5   20–5 Describe **two** risks associated with heart or heart-and-lung transplants. [2]

Total: 7

p53–4 4.2.2.2

21 **Figure 6** shows the structure of three types of blood vessel: an artery, a capillary and a vein.

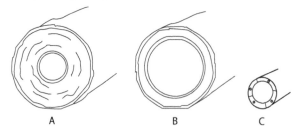

Figure 6

21–1 Identify the blood vessels labelled A–C. [3]

QWC 21–2 Describe the structures and functions of arteries, capillaries and veins. [6]

Total: 9

p55–6 4.2.2.2

22 Blood is a tissue. **Figure 7** shows the main components of the blood.

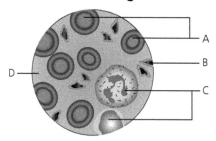

Figure 7

WS3.5 22–1 Name the components of the blood labelled A–D in **Figure 7**. [4]

22–2 Match each component of the blood, 1–4, to its function, A–D. [4]

Component of the blood	Function
1 Red blood cells	A Destroying pathogens
2 White blood cells	B Transporting substances around the body in solution
3 Platelets	C Transporting oxygen around the body
4 Plasma	D Helping blood to clot

22–3 The blood plasma transports carbon dioxide.

Name **two** other substances transported by the plasma. [2]

22–4 Red blood cells are adapted for their function.

Describe **one** adaptation of a red blood cell. [2]

Total: 12

23 Diseases can be communicable or non-communicable.

p59&79 4.2.2.5

23–1 What is the difference between communicable and non-communicable diseases? [1]

p59 4.2.2.5

23–2 Communicable and non-communicable diseases are major causes of ill health.

Give **two** other factors that can affect both physical and mental health. [2]

p59   4.2.2.5

23–3 Different types of disease may interact.

Match the health issues, 1–4, with the disease that the health issue most commonly triggers, A–D. [4]

Health issue
1 Immune system defects
2 Viruses living in cells
3 Severe physical ill health
4 Immune reactions caused by a pathogen

Disease most commonly triggered
A Certain cancers
B Infectious diseases
C Allergies, such as skin rashes and asthma
D Depression and other mental illness

p61   4.2.2.6

23–4 Risk factors are linked to an increased rate of a disease.

Explain the difference between **causation** and **correlation** with risk factors. [2]

p55, 60 &62   4.2.2.6

23–5 Obesity is a risk factor for type 2 diabetes.

People with type 2 diabetes produce enough insulin, but their cells are not sensitive to insulin.

Figure 8 shows the relationship between percentage abdominal fat and insulin sensitivity in a group of people.

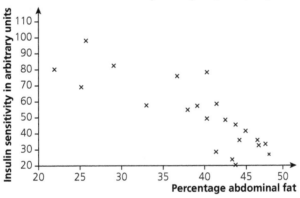

Figure 8

Name **two** other diseases that obesity is a risk factor for. [2]

p62&285   4.2.2.6   MS2g   WS3.5

23–6 Identify the pattern shown in the graph in **Figure 8**. [1]

p62   4.2.2.6   WS3.5, 3.6

23–7 A student uses the information from the graph in **Figure 8** to conclude that percentage abdominal fat can be used to find out if a person has type 2 diabetes. Give **two** reasons why this conclusion is not correct. [2]

Total: 14

p62   4.2.2.6

24   Drinking too much alcohol can affect a person's health.

24–1 Describe how a person's liver can be affected by drinking too much alcohol. [2]

24–2 Describe the short-term and long-term effects that alcohol can have on brain function. [2]

24–3 What can happen to unborn babies if they are exposed to high levels of alcohol? [1]

WS1.4

24–4 Excessive alcohol consumption can have wider effects on society.

Describe **two** human and **two** financial costs of excess alcohol consumption. [4]

Total: 9

p60–2 4.2.2.6

25 Carcinogens are chemicals or agents that can cause cancer by damaging DNA.

25–1 Give **two** examples of carcinogens. [2]

25–2 Cancer causes changes in cells that lead to tumours forming. Describe how tumours form. [1]

25–3 Tumours can be benign or malignant.

Give **one** difference between benign tumours and malignant tumours. [1]

25–4 Describe how tumours can spread to other parts of the body. [1]

25–5 Risk factors can increase the chance of a person getting cancer.

Risk factors can be caused by genetics or lifestyle factors.

List **three** lifestyle risk factors for cancers that scientists have identified. [3]

Total: 8

Plant tissues, organs and systems

Quick questions

p23& 69–71 4.1.2.3, 4.2.3.1

1 Where is meristem tissue found in plants and what is its function?

p71–2 4.2.3.2

2 Give the term for the evaporation of water from leaves and list **four** factors that increase the rate of evaporation.

p71–2 4.2.3.2

3 Describe what the transpiration stream is.

p72 4.2.3.2

4 Describe **two** functions of transpiration.

p68&71 4.2.3.2

5 Give the function of stomata and state where most stomata are found.

p68 4.2.3.2

6 Name the cells that control the size of the stomata.

p69&72 4.2.3.2

7 Describe what 'translocation' is.

Exam-style questions

8 Leaves are the main sites of photosynthesis in plants.

p71 4.2.3.1

8–1 What word correctly describes a leaf? Choose your answer from the options below. [1]

Cell	Organ	System	Tissue

p68–9 | 4.2.3.1, 4.2.3.2

8–2 **Figure 9** shows a cross-section through a leaf.

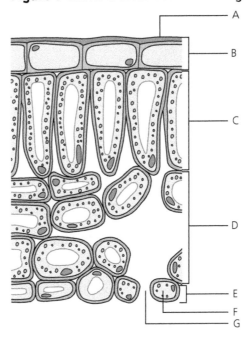

Figure 9

Name the parts of the leaf labelled A–G. Choose your answers
from the options below. *[7]*

cuticle	guard cell	lower epidermis	stoma
spongy mesophyll tissue	palisade mesophyll tissue	upper epidermis	

p68–69 | 4.2.3.1, 4.2.3.2

8–3 Match the plant tissues, 1–5, to their functions, A–E. *[5]*

Plant tissue
1 Epidermal
2 Palisade mesophyll
3 Spongy mesophyll
4 Xylem
5 Phloem

Functions
A Transport dissolved sugars from the leaves
B Cover and protect the plant
C Transport water and mineral ions from the roots
D Main site of photosynthesis
E Main site of gas exchange

p68–9 | 4.2.3.1, 4.2.3.2 | QWC

8–4 Describe how named plant tissues are adapted for their functions. *[6]*

Total: 19

9 Plants have an organ system for transport of substances.

p71 | 4.2.3.2

9–1 Name the **three** organs in the plant organ system that transports
substances around a plant. *[1]*

p71–2 | 4.2.3.2

9–2 Water and mineral ions are absorbed from the soil by root hair cells.

Give **one** use of water in plants. *[1]*

p36&70 | 4.1.1.2, 4.1.1.3, 4.2.3.2

9–3 Explain why root hair cells contain large numbers of mitochondria. *[4]*

p71–2 | 4.2.3.2

9–4 Water is transported from the roots to the leaves and lost from
the leaves by transpiration.

Describe the process of transpiration. *[4]*

p71–2 | 4.2.3.2

9–5 Explain the **four** environmental factors that would increase the
rate of transpiration. *[4]*

Total: 14

10 Stomata are found on the leaves of plants.

Figure 10 shows stomata on the lower epidermis of a leaf.

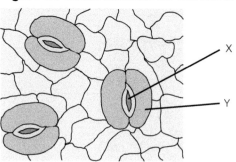

X

Y

Figure 10

10–1 Name the parts labelled X and Y in **Figure 10**. [2]

10–2 Plants can open and close their stomata.

Give the reason why plants close their stomata. [1]

WS3.5

10–3 **Figure 11** shows the appearance of open and closed stomata.

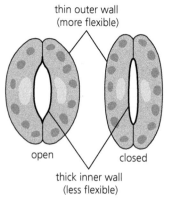

thin outer wall
(more flexible)

open   closed

thick inner wall
(less flexible)

Figure 11

Stomata open when the light intensity is high.

When the light intensity is high, potassium ions (K^+) are transported into guard cells.

Suggest why a high concentration of K^+ inside guard cells causes stomata to open. [4]

AT7
MS5c

10–4 A student investigates the distribution of stomata on the upper and lower surfaces of a leaf.

The student uses a microscope to look at the lower epidermis, and counts the number of stomata in one field of view.

Figure 12 shows the field of view.

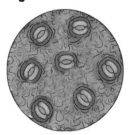

Figure 12

The diameter of the field of view is 0.4 mm.

Calculate the number of stomata in $1\,mm^2$ of the leaf.

Use $\pi = 3.14$

Give your answer to the nearest whole number. [3]

WS2.5, 2.7

10–5 The student counts the number of stomata in one field of view and uses this count to estimate the number of stomata in $1\,mm^2$ of the leaf.

Suggest how the student could improve their estimate. [2]

WS2.2

10–6 Suggest a method the student could use to estimate the area of a leaf. [2]

WS3.5

10–7 The student finds that there are more stomata on the lower surface than the upper surface.

Suggest why this is an advantage for a plant. [2]

Total: 16

Organisation topic review

1 The heart pumps blood around the body through the blood vessels.

p51 4.2.2.1

1–1 Copy and complete **Table 2** to show whether each structure is a tissue, organ or organ system.

Tick **one** box for each structure. [3]

Structure	Tissue	Organ	Organ system
Blood			
Blood, blood vessels and heart			
Heart			

Table 2

p55–6 4.2.2.3 WS2.6, 3.5 AT7

1–2 **Figure 13** shows some blood cells seen using a light microscope They are magnified ×2000.

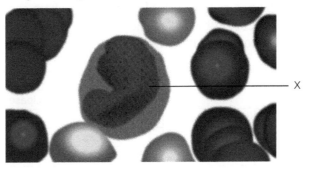

Figure 13

Draw a scientific drawing of cell **X** from **Figure 13**. Label **two** parts of the cell. [3]

p53–4 4.2.2.2

1–3 Arteries and veins are two types of blood vessel.

Compare the structure of an artery with the structure of a vein. *[3]*

4.2.2.2 WS3.3, 4.2,4.4, 4.5 / MS1c,2h

1–4 **Table 3** shows the speed of blood flow through different blood vessels.

Blood vessel	Speed of blood flow in cm³/s
Aorta	40
Capillary	0.03
Vena cava	15

Table 3

Calculate the volume of blood that flows through the aorta in 1 hour. Give your answer in dm^3 *[2]*

p51–3 4.2.2.2

1–5 What causes the high speed of blood flow in the aorta? *[1]*

4.2.2.2 MS1c,2a / WS3.2, 3.3,4.6

1–6 Calculate the percentage decrease in the speed of blood flow from the aorta to the capillaries.

Give your answer to 3 significant figures. *[2]*

p51–3 4.2.2.2

1–7 Explain the benefit of the slow rate of blood flow in a capillary. *[2]*

Total: 16

2 Enzymes are produced by different organs in the digestive system.

p42, 44 &48 4.2.2.1

2–1 Copy and complete the sentences. *[4]*

The digestive system has several organs that work together to digest and _____ food.

Digestive enzymes convert food into small _____ molecules that can enter the blood stream.

Enzymes catalyse specific reactions in living organisms due to the shape of their _____ _____.

Enzymes are _____ at high temperatures.

p43–4 4.2.2.1

2–2 Copy and complete **Table 4** to show where each enzyme is produced. Put a tick if the organ produces the enzyme. You can tick more than one enzyme for each organ. *[3]*

		Enzyme		
		Amylase	Lipase	Protease
Organ	Salivary gland			
	Stomach			
	Pancreas			

Table 4

p47 4.2.2.1 RP4

2–3 Amylase breaks down starch to sugars.

Describe how you would test a sample of food to show that it contains sugars. *[3]*

p45 4.2.2.1

2–4 Lipase is an enzyme that breaks down lipids (fats and oils).

Name the products of lipid digestion. *[1]*

p46 4.2.2.1

2–5 A person who has had their gall bladder removed may have a slower rate of lipid digestion. Explain why. *[3]*

2–6 Lifestyle factors are linked to an increased rate of obesity.

4.2.2.6 WS3.2 MS4a

The graph in **Figure 14** shows how the percentage of obese adults in the UK changed between 1994 and 2006.

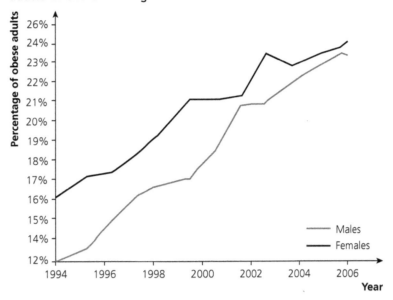

Figure 14

Describe the patterns shown in the graph in **Figure 14**.　　　*[3]*

p58–9 4.2.2.6 WS3.5

2–7 Suggest **two** possible reasons for the change in the percentage of obese adults.　　　*[2]*

Total: 19

p72 4.2.3.2

3 Plants lose water from their leaves by transpiration.

A student investigated water loss from a plant, using the apparatus shown **Figure 15**.

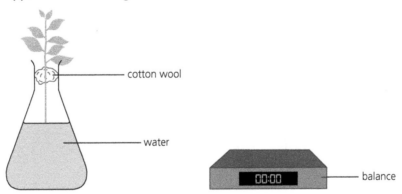

Figure 15

The student followed this method:

- Set up the flask with the plant, water and cotton wool.
- Record the starting mass of the flask.
- Record the mass of the flask every two hours.
- Calculate the mass of water lost by the plant.

Table 5 shows the student's results.

Time in hours	Total mass of water lost from the plant in grams
0	0.0
2	0.8
4	1.6
6	4.1
8	6.6
10	9.2
12	11.6

Table 5

WS2.3

3–1 Suggest why the student used cotton wool when they set up the apparatus. [1]

WS3.1, 3.2
MS4c

3–2 Plot a graph of the student's results. [4]

WS3.3
MS4d

3–3 The rate of water loss between 0 and 4 hours was different from the rate of water loss between 4 and 12 hours.

Calculate the rates of water loss between:

- 0 and 4 hours

- 4 and 12 hours.

Give a suitable unit for rate. [3]

WS3.5

3–4 Suggest **one** explanation for the difference in the two rates of water loss. [2]

Total: 10

p56–9 4.2.2.4, 4.2.2.6

4 A group of scientists researched the relationship between alcohol intake and the risk of coronary heart disease.

Figure 16 shows the scientists' results.

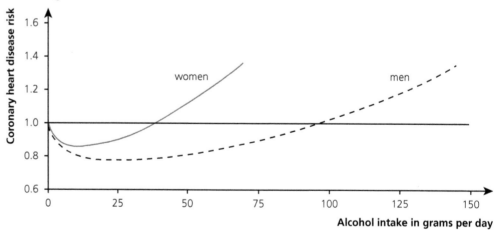

Figure 16

WS3.5

4–1 Use **Figure 16** to give **one** conclusion about the relationship between alcohol intake and the risk of coronary heart disease. [1]

WS3.7

4–2 A journalist uses the data to claim:

'Drinking an alcoholic drink every day is good for your health.'

Evaluate this claim. Use information from **Figure 16**. *[3]*

4–3 The rate of blood flow through the coronary artery of a healthy person is 250 cm^3 min^{-1}

Explain why the rate of blood flow through the coronary artery would be different in a person with coronary heart disease. *[3]*

QWC
WS1.3, 1.4,1.5

4–4 Statins are drugs that can be used to reduce the risk of serious heart conditions.

Read the information below.

Statins are a group of drugs that lower the level of low density lipoprotein (LDL) cholesterol in the blood. Having a high level of LDL in the blood can be dangerous as it can cause narrowing of the arteries. If the coronary arteries are narrowed then the blood supply to the heart can be restricted. Scientific studies have shown that 1 in 50 people who take statins for five years will avoid a serious condition, such as a heart attack or stroke.

Statins are tablets that are taken once every day. People taking statins will take them for the rest of their life as their blood cholesterol level will rise if they stop taking them. Statins can interact with other medicines leading to side effects. Some people experience minor side effects such as headache, diarrhoea and feeling sick. Rarely, people can suffer more serious side effects like muscle or kidney damage. Blood cholesterol level can also be reduced by eating a healthy diet, exercising regularly, not smoking, and reducing alcohol intake.

Evaluate the use of statins to reduce the risk of serious heart conditions. *[6]*

Total: 13

3 Infection and response

Communicable disease

Quick questions

p79   4.3.1.1   1   What are pathogens?

p79   4.3.1.1   2   What types of organisms can be pathogens?

p81   4.3.1.2   3   What type of pathogen causes measles?

p82   4.3.1.2   4   Which body cells are attacked by HIV?

p82   4.3.1.2   5   Which group of drugs are used to control HIV?

p103   4.3.1.2   6   What type of organisms are affected by tobacco mosaic virus?

p82   4.3.1.3   7   What type of pathogen causes salmonella food poisoning?

p82–3   4.3.1.3   8   What type of pathogen causes gonorrhoea?

p103   4.3.1.4   9   Describe the effects of rose black spot.

p103   4.3.1.4   10   What type of pathogen causes rose black spot?

p84–5   4.3.1.5   11   What type of pathogen causes malaria?

p87   4.3.1.6   12   Which system destroys pathogens that enter the body?

p90   4.3.1.8   13   What are antibiotics?

p92   4.3.1.8   14   Aspirin is a type of medicine called a painkiller. What is the purpose of a painkiller?

p90   4.3.1.9   15   Name the scientist who discovered penicillin.

p93   4.3.1.9   16   What is a 'placebo'?

p93   4.3.1.9   17   What is a 'double blind trial'?

Exam-style questions

18   Pathogens can spread between plants or between animals.

p80–1   4.3.1.1   18–1 Describe **three** ways that pathogens can spread.   *[3]*

p85–7   4.3.1.6   18–2 The human body has defence systems against pathogens. Describe how each of the systems listed below defends against pathogens.   *[4]*

- Skin

- Hairs in nose

- Ciliated epithelium in trachea and bronchi

- Hydrochloric acid in the stomach

p87–8   4.3.1.6   QWC   18–3 If pathogens enter the body, white blood cells defend the body.

Explain how the white blood cells defend the body against pathogens.   *[6]*

Total: 13

p82   4.3.1.1

19   Bacteria are pathogens that can cause disease.

19–1 Describe how bacteria can make us feel ill.   *[2]*

19–2 Salmonella is a type of food poisoning caused by bacteria.

Give **one** way that salmonella is spread.   *[1]*

19–3 Give **two** symptoms of salmonella food poisoning.   *[2]*

19–4 Explain what causes the symptoms of salmonella food poisoning.   *[2]*

19–5 Suggest **one** way of controlling the spread of salmonella.   *[1]*

Total: 8

20   Gonorrhoea is a disease caused by bacteria.

p82–3   4.3.1.3   20–1 How is gonorrhoea spread?   *[1]*

p82–3   4.3.1.3   20–2 What are the symptoms of gonorrhoea?   *[1]*

p82–3   4.3.1.3   20–3 How can the spread of gonorrhoea be controlled?   *[1]*

p90   4.3.1.8   20–4 Antibiotics have greatly reduced the number of deaths from infectious diseases.

Explain why it is important to use specific antibiotics to treat a bacterial infection.   *[2]*

p82–3 &91   4.3.1.3, 4.3.1.8   20–5 Why can gonorrhoea not be treated with penicillin anymore?   *[1]*

Total: 6

21   Viruses are pathogens that can cause disease.

p81   4.3.1.1   21–1 Describe how viruses can make us ill.   *[2]*

p82   4.3.1.2   21–2 HIV is a virus. How is HIV spread?   *[1]*

p82   4.3.1.2   21–3 What are the initial symptoms of HIV infection?   *[1]*

p82   4.3.1.2   21–4 What happens in late-stage HIV infection or AIDS?   *[1]*

p90   4.3.1.8   21–5 Antibiotics cannot be used to treat viral diseases. Give **one** reason why.   *[1]*

pXX   4.3.1.1, 4.3.1.8   21–6 Explain why it is difficult to develop drugs that kill viruses.   *[2]*

Total: 8

22   Measles is a disease caused by a virus.

p81   4.3.1.2   22–1 Give **two** symptoms of measles.   *[2]*

p81   4.3.1.2   22–2 Describe how measles is transmitted (spread).   *[2]*

p89   4.3.1.6   22–3 Young children are vaccinated against measles. What is a 'vaccine'?   *[1]*

p81   4.3.1.2   22–4 Why is it important that babies are vaccinated against measles?   *[1]*

p89   4.3.1.7   QWC   22–5 Explain how a vaccine can prevent measles.   *[6]*

Total: 12

23 Tobacco mosaic virus (TMV) and rose black spot are both plant diseases.

p103   4.3.1.2

23–1 What are the symptoms of TMV? [1]

p103   4.3.1.4

23–2 What are the symptoms of rose black spot? [1]

p103   4.3.1.2, 4.3.1.4

23–3 Explain why TMV and rose black spot both affect plant growth. [3]

p103   4.3.1.4

23–4 Describe **two** ways that rose black spot can be treated. [2]

Total: 7

p84–5   4.3.1.5

24 Malaria is a disease caused by a protist.

24–1 How is malaria spread? [1]

24–2 What effects does malaria have on the body? [2]

24–3 Explain **two** methods of controlling the spread of malaria. [4]

Total: 7

p90–3   4.3.1.9

25 Some drugs were traditionally extracted from plants and microorganisms.

25–1 Match the drugs, 1–3, to the organism it was originally extracted from, A–D. [3]

Name of drug	**Organism the drug was extracted from**
1 Aspirin	A Mould
2 Digitalis	B Willow tree
3 Penicillin	C Foxglove
	D Tobacco plant

25–2 Digitalis is a heart drug. Name the group of drugs that:

- aspirin belongs to [1]
- penicillin belongs to. [1]

25–3 Describe the difference between the traditional source of drugs and the source of new drugs. [2]

25–4 New drugs are tested extensively for toxicity, efficacy and dose. Give the reasons why drugs are tested for each of these. [3]

Total: 10

p93   4.3.1.9

26 Here are the main stages when testing a new drug:

- research and development
- pre-clinical testing
- clinical trials.

26–1 Pre-clinical testing of a new drug happens in a laboratory.

What are drugs tested on during pre-clinical testing? [1]

26–2 What is the main purpose of pre-clinical testing? [1]

26–3 There are three phases of clinical trials: phase 1, phase 2 and phase 3. During these trials, drugs are tested on healthy volunteers and patients.

At the start of clinical trials, phase 1 uses small numbers of healthy volunteers.

Give the reason why very low doses of drug are given at the start of these phase 1 clinical trials. [1]

26–4 Why do phase 1 trials use healthy volunteers rather than patients with the disease? [1]

26–5 Phase 2 and phase 3 are further clinical trials that use large numbers of patients.

What is the main purpose of these clinical trials? [1]

26–6 Some clinical trials are double blind trials with one group getting the drug and one group getting a placebo.

Why does one group get a placebo in a double blind trial? [1]

26–7 The results of testing and trials are only published after scrutiny by peer review.

What does peer review mean? [1]

Total: 7

ⒽMonoclonal antibodies

Quick questions

p98 | 4.3.2.1 | 1 What are monoclonal antibodies?

p87 | 4.3.2.1 | 2 What type of cell makes antibodies?

p87&98 | 4.3.2.1 | 3 How does the body respond to foreign antigens?

p98 | 4.3.2.1 | 4 What type of cell does a lymphocyte need to be combined with to make a cell that produces monoclonal antibodies?

p98 | 4.3.2.1 | 5 What type of cell produces monoclonal antibodies?

Exam-style questions

p98–100 | 4.3.2.1 | 6 Monoclonal antibodies can be made in a laboratory.

6–1 Copy and complete the description of monoclonal antibodies. Choose your answers from the options below. [5]

antibody	antigen	chemical	site	specific	target

Monoclonal antibodies are _____ to one binding _____ on one protein _____ and so are able to _____ a specific _____ or specific cells in the body.

6–2 **Figure 1** shows the process of producing monoclonal antibodies.

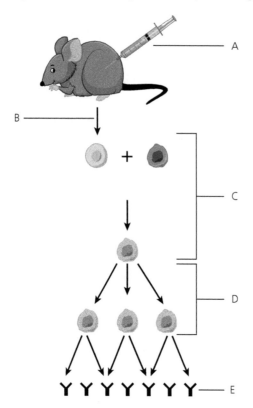

Figure 1

Copy and complete the flowchart below to describe how
monoclonal antibodies are produced. [5]

A	A specific _____ is injected into a mouse.

↓

B	The mouse _____ produce antibodies specific to the antigen.

↓

C	The lymphocytes are combined with a tumour cell to make a _____ cell.

↓

D	The hybridoma cell is _____ producing many identical cells that make the same antibody.

↓

E	A large amount of _____can be collected and purified.

6–3 Monoclonal antibodies can be used in diagnostic tests and to treat disease.

Describe **two** ways that monoclonal antibodies can be used in
diagnostic tests. [2]

WS1.4
QWC

6-4 When a woman is pregnant, a hormone called human chorionic gonadotrophin (HCG) is present in her urine. Pregnancy tests use monoclonal antibodies to detect the HCG.

Figure 2 shows the structure of a pregnancy test.

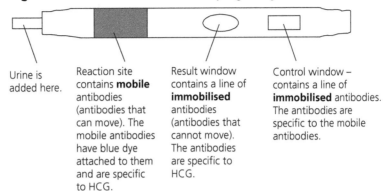

Urine is added here.

Reaction site contains **mobile** antibodies (antibodies that can move). The mobile antibodies have blue dye attached to them and are specific to HCG.

Result window contains a line of **immobilised** antibodies (antibodies that cannot move). The antibodies are specific to HCG.

Control window – contains a line of **immobilised** antibodies. The antibodies are specific to the mobile antibodies.

Figure 2

Explain how a pregnancy test can be used to show that a woman is pregnant. Use information from **Figure 2** and your knowledge of monoclonal antibodies. [6]

Total: 18

p100 **4.3.2.2**

7 Monoclonal antibodies (MABs) can be used for targeted drug therapy to treat cancer. MABs work by recognising and finding specific proteins on cancer cells.

Herceptin is a targeted cancer drug used as a treatment for cancers including early breast cancer, advanced breast cancer and advanced stomach cancer. Some breast and stomach cancers have large amounts of a protein called human epidermal growth factor receptor 2 (HER2).

HER2 makes the cancer cells grow and divide. Herceptin works by attaching to HER2 so it stops the cancer cells from growing and dividing.

Herceptin can cause many side effects including allergic reactions, joint and muscle pain, diarrhoea, heart problems and fatigue (tiredness).

7-1 Explain how MABs can be used to **treat** cancer. [2]

WS1.4
QWC

7-2 Scientists can use mice to make MABs in a laboratory. Each MAB recognises one specific protein, so different MABs have to be made to target different types of cancer.

Describe how scientists produce MABs to treat breast cancer. [6]

7-3 Explain how MABs, such as Herceptin, stop cancer cells from growing and dividing.

Use information from the text above and your own knowledge of monoclonal antibodies. [3]

7-4 Evaluate the use of MABs such as Herceptin to treat cancer.

Use information from the text above and your own knowledge of monoclonal antibodies. [4]

7-5 Give the reason why monoclonal antibodies are not yet as widely used as everyone hoped when they were first developed. [1]

Total: 16

Plant diseases

Quick questions

p103 4.3.1.2, 4.3.3.1
1 What type of pathogen causes tobacco mosaic virus (TMV)?

p103 4.3.1.4, 4.3.3.1
2 What type of pathogen causes rose black spot?

p104 4.3.3.1
3 Name **one** type of insect pest that can affect plants.

p105 4.3.3.1
4 Which ions are needed for protein synthesis?

p105 4.3.3.1
5 Which ions are needed to make chlorophyll?

Exam-style questions

p105 4.3.3.1
6 Plants can be infected by pathogens and damaged by ion deficiencies.

6–1 Give **three** ways that plant disease can be detected. [3]

WS1.4 6–2 Describe **three** ways that a gardener could identify a plant disease. [3]

WS1.4 6–3 Give the reason why horticulturalists (people who grow plants) need to know about ion deficiencies. [1]

6–4 Describe the appearance of a plant with:

• magnesium ion deficiency [1]

• nitrate ion deficiency. [1]

6–5 Plants deficient in magnesium ions may have poor growth. Explain why. [5]

Total: 14

p106 4.3.3.1, 4.3.3.2
7 Plants have defence mechanisms to protect them from microorganisms and from animals.

7–1 Describe **three** physical defence responses that plants have to resist the invasion of microorganisms. [3]

7–2 Describe **two** chemical defence responses that plants use against other organisms. [2]

7–3 Plants can have mechanical adaptations that increase their chance of survival. Suggest why it is an advantage for a plant to have leaves that droop or curl when they are touched. [1]

7–4 Some plants use mimicry to trick animals.

The leaves of the passion flower plant have yellow spots on them as shown in **Figure 3.**

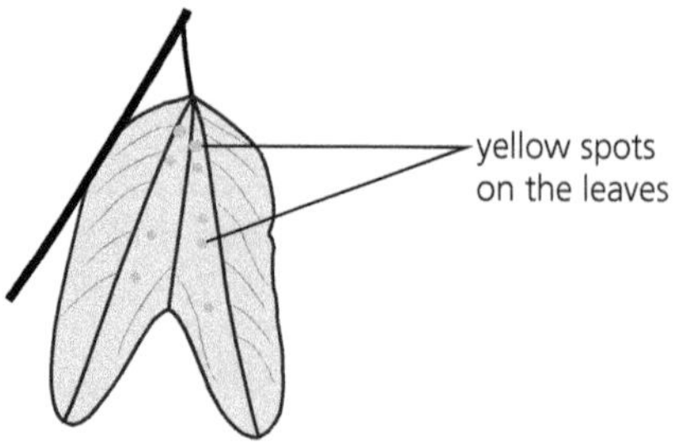

yellow spots on the leaves

Figure 3

Female butterflies are less likely to lay their eggs on plants that already have butterfly eggs.

Suggest why these spots are an advantage for the plant. [2]

7–5 Some plants have thorns on their stems or stinging hairs on their stems and leaves.

The raspberry plant has prickles on its stem as shown in **Figure 4.**

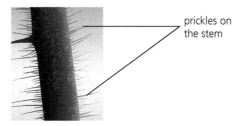

prickles on the stem

Figure 4

Explain how the prickles help the plant to survive. *[2]*

Total: 10

Infection and response topic review

1 Measles is a viral disease that most young children are vaccinated against.

p86 4.3.1.6

1–1 Describe how the non-specific defence systems of the human body try to prevent the entry of the measles virus. *[3]*

p89 4.3.1.7

1–2 Not everyone can be vaccinated against measles.

Why is it an advantage if a large proportion of the population is vaccinated against measles? *[1]*

p89 4.3.1.7 WS3.5

1–3 Young children have their first measles vaccination at around the age of one year, and then a second measles vaccination when they are three years old.

The graph in **Figure 5** shows the number of antibodies against the measles virus in a child's blood after the first and second measles vaccine.

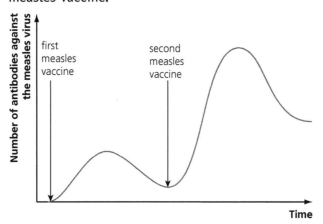

Figure 5

Describe **two** differences in the production of antibodies after the second vaccination compared with after the first vaccination. *[2]*

p89 4.3.1.2, 4.3.1.7 WS1.5

1–4 Measles is a serious disease that can be fatal if complications arise. Complications include:

- 1 in 200 chance of seizures (fits)
- 1 in 16 chance of pneumonia
- 1 in 1000 chance of brain swelling that can cause brain damage
- 1 in 5000 chance of death.

The measles vaccine can cause side effects including a rash, high temperature, and feeling unwell for two or three days. Some rare side effects of the vaccine include:

- 1 in 1000 chance of seizures (fits)
- 1 in 900 000 chance of severe allergic reaction.

Some people are concerned about having their children vaccinated due to the risk of side effects.

Evaluate the use of the measles vaccine. [4]

Total: 10

2 Tomato spotted wilt virus (TSWV) is a plant disease that is transmitted by small insects called thrips.

TSWV causes yellow patches on leaves due to a lack of chlorophyll, and stunted growth.

p103& 105 4.3.1.2, 4.3.3.1

2–1 Explain why plants infected with TSMV have stunted growth. [4]

p98–9 4.3.2.1, 4.3.3.1

(H) 2–2 It is difficult to diagnose TSWV by its symptoms as other plant pathogens cause similar symptoms.

A scientist wants to develop a diagnostic testing kit for TSWV using monoclonal antibodies.

The scientist isolates an antigen from the surface of the virus that causes TSWV.

Describe how the scientist uses the antigen to produce monoclonal antibodies against TSWV. [4]

p106 4.3.3.2

2–3 Some plants produce oils that they secrete onto the surface of their leaves.

Suggest **one** reason why oils on leaves can protect plants from insects, such as thrips. [1]

p104 4.3.3.1

2–4 Suggest **one** way that a tomato grower could reduce the spread of TSWV. [1]

Total: 10

3 Salmonella is a bacterial disease that causes symptoms including abdominal cramps, diarrhoea and vomiting.

p82&88 4.3.1.1, 4.3.1.3

3–1 What causes the symptoms of salmonella? [1]

p81&90 4.3.1.1, 4.3.1.8

3–2 Many bacterial diseases can be treated with antibiotics, but viral diseases cannot.

Explain why antibiotics cannot be used to treat viral diseases. [2]

p82& 87–88 4.3.1.2, 4.3.1.6

3–3 Explain why a person with late stage HIV infection may take longer to recover from a salmonella infection than a healthy person. [2]

p82&93 4.3.1.2, 4.3.1.9 QWC

3–4 Scientists develop a new antiretroviral drug to treat HIV.

Explain the testing that this new drug must undergo before it can be used to treat people. [6]

Total: 11

4 Bioenergetics

Photosynthesis

Quick questions

p111   4.4.1.1

1   Explain photosynthesis in terms of energy transfer.

p111–12   4.4.1.1

2   What is an 'endothermic reaction'?

p111   4.1.1.2, 4.4.1.1

3   Name the green pigment found in chloroplasts that absorbs light energy.

p111   4.4.1.1

4   Write the word equation for photosynthesis.

p111   4.4.1.1

5   Give the chemical symbols for carbon dioxide, glucose, oxygen and water.

p112   4.4.1.2

6   What is a 'limiting factor'?

p112   4.4.1.2

7   List the four main limiting factors for photosynthesis.

Exam-style questions

8   Photosynthesis requires light energy.

p111   4.4.1.1

8–1   What type of chemical reaction is photosynthesis?   *[1]*

p111   4.4.1.1

8–2   Copy and complete the balanced symbol equation for photosynthesis.   *[2]*

$$6CO_2 + 6\underline{\hspace{1.5cm}} \xrightarrow{\text{light}} \underline{\hspace{1.5cm}} + 6O_2$$

p115   4.4.1.3a

8–3   Photosynthesis produces glucose that can be used to produce amino acids for protein synthesis. Describe **four** other ways that plants use glucose.   *[4]*

p113–14   4.4.1.2   RP6   AT1,3,4,5   QWC

8–4   The rate of photosynthesis can be affected by light intensity.

Figure 1 shows some apparatus used to investigate photosynthesis.

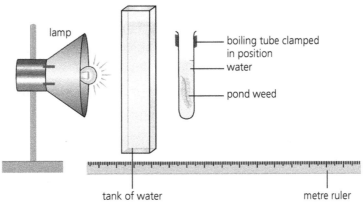

Figure 1

Describe how you would use the apparatus to investigate the effect of light intensity on the rate of photosynthesis.

You should include details of the measurements you would take and how you would carry out a fair test. [7]

Total: 14

9 A student investigated the effect of light intensity on the rate of photosynthesis by counting the number of bubbles produced by pondweed.

The student took measurements at distances between the pondweed and the lamp of 10, 20 and 40 cm.

Table 1 shows the student's results.

Distance between lamp and pondweed in cm	Number of bubbles released in 1 minute
10	61
20	15
40	4

Table 1

9–1 Suggest **one** way the student could improve their results. [1]

9–2 Explain why more bubbles are released at 10 cm than at 20 cm. [2]

9–3 Each bubble released by the pondweed has a diameter of 1 mm.

The volume of a bubble can be calculated using:

$$V = \frac{4}{3}\pi r^3$$

where r = radius and π = 3.14

Calculate the rate of gas production in mm^3 per minute when the pondweed is 40 cm from the lamp. [4]

9–4 Light intensity can be calculated using the inverse square law:

$$\text{light intensity} \propto \frac{1}{\text{distance squared}}$$

Explain how doubling the distance between the lamp and the pondweed affects **light intensity**. Include calculations in your answer. [3]

9–5 The student repeats the investigation at a distance of 80 cm. Predict the number of bubbles released in 1 minute. Give a reason for your prediction. [2]

9–6 The student repeats the investigation at a distance of 5 cm and counts 60 bubbles. Explain why the number of bubbles is similar at 5 cm and 10 cm. [2]

9–7 The rate of photosynthesis can be affected by carbon dioxide concentration and temperature. Describe and explain how these **two** factors affect the rate of photosynthesis. You may include sketch graphs in your answer. [6]

Total: 20

p49 | 4.4.1.2 | RP5
p112–113 | 4.4.1.2
4.4.1.2 | MS3d
p114 | 4.4.1.2 | MS3a, 3d
p49 | 4.4.1.2 | RP5
p49 | 4.4.1.2
p48–9 & 112–13 | 4.2.2.1, 4.4.1.2 | QWC

p112–13   4.4.1.2

10   Plants can be grown in greenhouses to improve the growth rate.

The graph in **Figure 2** shows the effect of carbon dioxide concentration on the rate of photosynthesis in tomato plants at two different temperatures.

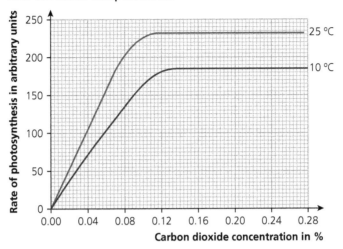

Figure 2

10–1 A student concludes that carbon dioxide concentration is not the limiting factor for photosynthesis when the temperature is 10°C and the carbon dioxide concentration is 0.16%.

Give the evidence for the student's conclusion from the graph.   *[1]*

H 10–2 State **two** factors that could be limiting the rate of photosynthesis when the temperature is 10°C and the carbon dioxide concentration is 0.16%.   *[2]*

H 10–3 Explain how a tomato grower could use the information from the graph to provide the optimum conditions for photosynthesis in a greenhouse.   *[4]*

H 10–4 Explain **two** reasons why the tomato grower does not use a carbon dioxide concentration above 0.12%.   *[2]*

Total: 9

Respiration

Quick questions

p120–1   4.4.2.1

1   Write a definition of cellular respiration.

p120   4.1.1.2, 4.4.2.1

2   Name the part of a cell where aerobic respiration happens.

p120& 123   4.4.2.1

3   What is the difference between aerobic and anaerobic respiration?

p120   4.4.2.1

4   Write the word equation for **aerobic** respiration.

p123   4.4.2.1

5   Write the word equation for **anaerobic** respiration in **muscles**.

p125   4.4.2.1

6   Write the word equation for **anaerobic** respiration in **plants and yeast**.

p125   4.4.2.1

7   What is another word for anaerobic respiration in yeast cells?

p125   4.4.2.3

8   What is 'metabolism'?

Exam-style questions

p120 4.4.2.1
p122 4.4.2.1
p123 4.4.2.1
p120& 123 4.4.2.1
p123 4.4.2.2
p124 4.4.2.2
p124 4.4.2.2
p123 4.4.2.2
p126 4.4.2.3

9   Aerobic and anaerobic respiration both use glucose and both transfer energy.

9–1 Give the chemical symbol for glucose. [1]

9–2 Give **two** reasons living things need energy. [2]

9–3 Give the reason why anaerobic respiration transfers much less energy than aerobic respiration. [1]

9–4 Anaerobic respiration transfers less energy than aerobic respiration. Describe **two** other differences between aerobic and anaerobic respiration in humans. [2]

9–5 Explain why a person cannot exercise efficiently when anaerobic respiration is taking place in muscle cells. [2]

H 9–6 An athlete completes a 100 metre sprint. The athlete's heart rate and breathing rate remain high after the sprint because the athlete has an oxygen debt.

Explain what is meant by an 'oxygen debt'. [2]

H 9–7 Describe what happens to the lactic acid that builds up during anaerobic respiration in muscles. [2]

Total: 12

10–1 During exercise, the heart rate increases to supply the muscles with more oxygenated blood. Give **two** other ways that the human body responds to the increased demand for oxygen in the muscles. [2]

10–2 Explain how an increase in heart rate helps a person to exercise. [3]

AT4
WS2.2
QWC

10–3 A student makes this hypothesis about the effect of exercise on heart rate:

'The heart rate of people who do not exercise regularly will take longer to return to normal after exercise than the heart rate of people who do regular exercise.'

Design an investigation to test this hypothesis. [6]

Total: 11

11   Metabolism is the sum of all the reactions in the body.

Metabolism includes both synthesis and breakdown of molecules in the body.

11–1 Match the smaller molecules, 1–3, to the larger molecules that can be synthesised, A–C. [3]

Smaller molecules	Larger molecules that can be synthesised
1 Glucose	A Glycogen
2 Glucose and nitrate ions	B Lipid
3 Glycerol and three fatty acids	C Amino acids and proteins

11–2 Name the molecule produced when excess proteins and amino acids are broken down. [1]

WS3.2 11–3 **Figure 3** shows the metabolic rate at different ages for males and females.

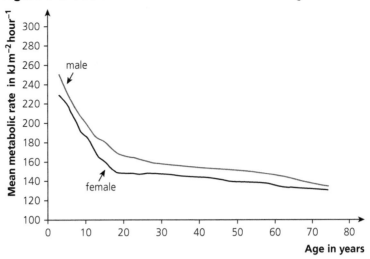

Figure 3

Compare the mean metabolic rates of females and males at different ages. Use information from **Figure 3**. [3]

MS1c 11–4 In males, the mean metabolic rate decreases from $205 \, kJ \, m^{-2} \, hour^{-1}$ at age 10 to $169 \, kJ \, m^{-2} \, hour^{-1}$ at age 20.

Calculate the percentage decrease in mean metabolic rate in males from 10 to 20 years of age. Give your answer to 3 significant figures. [3]

WS3.5 11–5 Suggest **one** reason why the metabolic rate is highest under 20 years of age. [3]

Total: 13

p125& 129   4.4.2.1   12   Yeast cells carry out anaerobic respiration.

Figure 4 shows the apparatus a student used to investigate the effect of temperature on the rate of aerobic respiration in yeast.

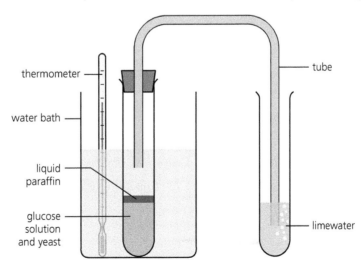

Figure 4

12–1 Give **one** use of anaerobic respiration in yeast in the manufacture of food and drinks. [1]

12–2 The student saw bubbles of gas come out of the tube and into the limewater. Name the gas found in the bubbles. *[1]*

WS2.2 12–3 Explain why the student added a layer of liquid paraffin. *[2]*

WS2.2 12–4 Suggest how the student could use the apparatus to measure the rate of respiration. *[2]*

WS2.2 12–5 The student used the same equipment and repeated the investigation at five different temperatures. Suggest **two** control variables the student used. *[2]*

Total: 8

Bioenergetics topic review

p111&
120–1 4.4.2.1

1 A student used a simple respirometer to investigate the rate of respiration in a grasshopper. **Figure 5** shows the equipment the student used.

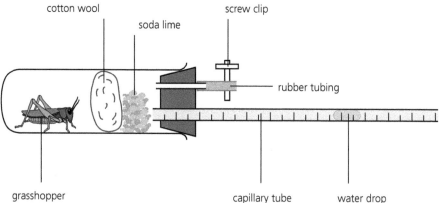

Figure 5

1–1 Name the gas produced by the grasshopper during this investigation. *[1]*

WS3.5 1–2 The soda lime in the tube absorbs carbon dioxide from the air.

Explain why the water drop moves towards the grasshopper during the investigation. *[3]*

WS2.4
AT4 1–3 Suggest why there is cotton wool between the grasshopper and the soda lime. *[1]*

WS2.3 1–4 Suggest the purpose of the screw clip. *[1]*

1–5 The student uses the same equipment to measure the rate of respiration in a plant, but covers the tube in black paper.

Suggest a reason for this using your knowledge of photosynthesis and respiration. *[3]*

Total: 9

p114 | 4.4.1.1, 4.4.2.1 | WS3.5 |

2 The graph in **Figure 6** shows the changes in carbon dioxide concentration in a greenhouse over 24 hours.

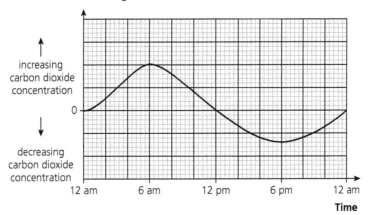

Figure 6

2–1 Explain why the carbon dioxide concentration rises between the times of 12 am and 6 am. [2]

2–2 Explain why there is a decrease in carbon dioxide concentration from 6 am to 6 pm. [2]

2–3 Suggest why there is no overall increase or decrease in the concentration of carbon dioxide over 24 hours. [2]

Total: 6

p123–6 | 4.4.2.2, 4.4.2.3 |

3 The energy transferred in respiration supplies all of the energy needed for living processes.

A student investigated the effect of exercise on their heart rate. The student measures their heart rate before, during and after running on a treadmill.

The graph in **Figure 7** shows the student's results.

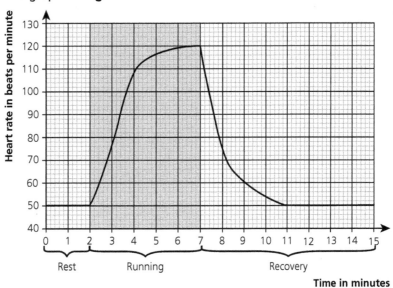

Figure 7

3–1 Give **two** examples of chemical reactions in humans to synthesise larger molecules. [2]

MS4a
WS3.2

3–2 Use the graph in **Figure 7** to find the student's maximum heart rate. *[1]*

MS4a
WS3.2

3–3 How long did it take for the student's heart rate to return to its resting rate after finishing exercise? *[1]*

3–4 Explain how the changes to the student's heart rate helped them to exercise. *[4]*

WS2.2

3–5 The student wants to modify the investigation to compare the time taken for the heart rate to return to normal in smokers and non-smokers. Suggest **two** variables the student would need to control so they could compare the results. *[2]*

Total: 10

5 Homeostasis and response

Homeostasis

Quick questions

p132 4.5.1

1   List **three** conditions in the human body that are controlled by homeostasis.

p132 4.5.1

2   What are the **two** main types of responses of the automatic control systems in the body?

p132 4.5.1

3   Name the type of cell that detects stimuli.

Exam-style questions

p132 4.5.1

4–1   Define the term 'homeostasis'.   [2]

p132 4.5.1 QWC

4–2   The control systems of the body can involve nervous or chemical responses. Describe the main features that are found in all control systems.   [6]

p132&
143 4.5.1

4–3   One of the conditions controlled in the human body is temperature. Explain why it is important to control body temperature.   [2]

p132&
151 4.5.1

4–4   Explain the importance of controlling blood glucose concentration.   [2]

p143–5&
156–7 4.5.1

4–5   **Figure 1** shows some organs of the human body.

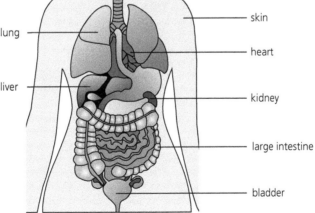

Figure 1

Name the organ that

• produces urea   [1]

• produces urine   [1]

• controls body temperature.   [1]

Total: 15

The human nervous system

Quick questions

p133 4.5.2.1
1 Name the **two** organs that make up the central nervous system (CNS).

p134 4.5.2.1
2 Name the type of cells that information passes along to reach the CNS.

p133 4.5.2.1
3 How does information pass along cells to reach the CNS?

p138 4.5.2.2
4 Name the organ that controls complex behaviour.

p141 4.5.2.3
5 The eye is a sense organ containing two types of receptors. What are the receptors sensitive to?

p141 4.5.2.3
6 Name the process of changing the shape of the lens in the eye to focus on near or distant objects.

p142 4.5.2.3
7 What are the common names of the eye defects myopia and hyperopia?

p143 4.5.2.4
8 Name the part of the brain that monitors and controls body temperature.

p144 4.5.2.4
9 What is the difference between vasoconstriction and vasodilation?

p144 4.5.2.4
10 What happens to skeletal muscles when we shiver?

Exam-style questions

p133–6 4.5.2.1
11 This question is about nervous coordination.

11–1 State the **two** main functions of the nervous system. *[2]*

11–2 When a person touches a sharp object, they quickly move their hand away.

Figure 2 shows the structure of the reflex arc that causes this response.

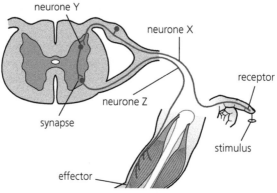

Figure 2

Give **two** other examples of reflex actions. *[2]*

11–3 Reflex actions are automatic, rapid actions that do not involve the conscious part of the brain. Why are reflex actions important? *[1]*

11–4 **Figure 2** shows a synapse. Explain what a synapse is. *[2]*

11–5 Describe how an impulse passes from neurone Y to neurone Z. *[3]*

QWC

11–6 Describe how the structures in a reflex arc cause the person's hand to move away from the sharp object. You should include the names of different types of neurones in your description. *[6]*

Total: 16

p135–7 4.5.2.1

12 Three students investigate their reaction times.

The students use this method:

- Student A sits with their lower arm resting on a table.
- Student B holds a ruler between student A's thumb and index finger.
- Student B drops the ruler and student A catches it.
- The distance on the ruler is recorded and a chart is used to convert the distance into a reaction time.
- The method is repeated 10 times and the mean calculated.
- The method is repeated to find the mean reaction times for students B and C.
- The students drink a cup of coffee and repeat the investigation.

Table 1 shows the mean reaction time for each student before and after drinking coffee.

Student	Mean reaction time before drinking coffee in milliseconds	Mean reaction time after drinking coffee in milliseconds
A	284	265
B	273	274
C	256	249

Table 1

12–1 Give the reason why the reaction the students investigate is not a reflex action. [1]

MS1c
WS4.4, 4.5

12–2 Give the reaction time before drinking coffee for student A in seconds in standard form. [1]

RP7
WS3.7

12–3 Explain why the students decided to measure their reaction times 10 times rather than three times. [2]

RP7
AT3
WS2.7

12–4 Suggest **two** improvements to the students' method to give valid results. [2]

RP7
WS2.7

12–5 The students use a computer program to test their reaction times. The students have to click a button when they see a red stop sign.

Suggest **two** reasons why measuring reaction times with a computer will give more valid results than the method the students used. [2]

RP7
WS3.5

12–6 Student C concludes that caffeine decreases reaction times. Evaluate this conclusion. [3]

Total: 11

p138–9 4.5.2.2

13 **Figure 3** shows a cross-section through the human brain.

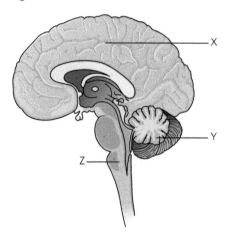

X

Y

Z

Figure 3

13–1 The brain is made of billions of interconnected cells. Name these cells. *[1]*

13–2 Name the structures labelled X, Y and Z in **Figure 3**. *[3]*

13–3 Describe the functions of structures X, Y and Z in **Figure 3**. *[3]*

Ⓗ 13–4 Give **two** reasons why investigating and treating brain disorders is very difficult. *[2]*

Ⓗ 13–5 Neuroscientists have been able to map the regions of the brain to particular functions by studying patients with brain damage.

Describe **two** other techniques that have been used to map brain function. *[2]*

WS1.5 Ⓗ 13–6 Give **one** advantage and **one** disadvantage of using patients with brain damage to investigate brain function. *[2]*

Total: 13

p140–2 4.5.2.3

14 **Figure 4** shows the structure of the human eye.

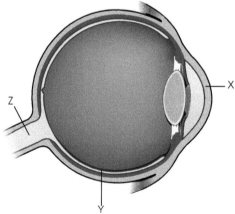

Z

X

Y

Figure 4

14–1 Identify the parts of the eye labelled X, Y and Z in **Figure 4**. *[3]*

14–2 Give the function of the sclera. *[1]*

14–3 A student enters a classroom where the light is dim. The student's eye changes in response to the change in light levels. Explain how the iris brings about the change. *[3]*

QWC 14–4 The student is reading a book and then looks up to read information the teacher has written on the board. Describe the changes that take place in the student's eye to allow them to focus on the distant object. [6]

WS1.4 14–5 Some people have myopia (short-sightedness) and wear spectacles (glasses) to correct their vision. **Figure 5** shows how a spectacle lens can correct myopia.

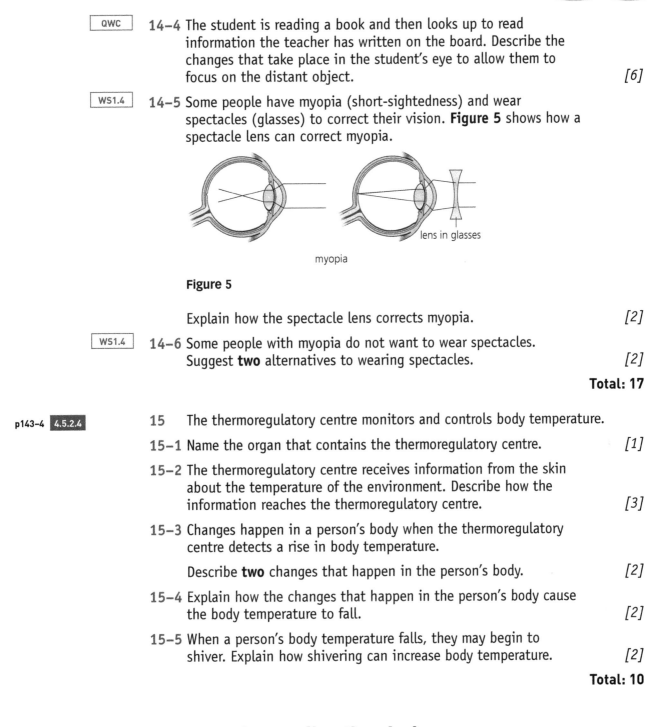

myopia

lens in glasses

Figure 5

Explain how the spectacle lens corrects myopia. [2]

WS1.4 14–6 Some people with myopia do not want to wear spectacles. Suggest **two** alternatives to wearing spectacles. [2]

Total: 17

p143–4 4.5.2.4 15 The thermoregulatory centre monitors and controls body temperature.

15–1 Name the organ that contains the thermoregulatory centre. [1]

15–2 The thermoregulatory centre receives information from the skin about the temperature of the environment. Describe how the information reaches the thermoregulatory centre. [3]

15–3 Changes happen in a person's body when the thermoregulatory centre detects a rise in body temperature.

Describe **two** changes that happen in the person's body. [2]

15–4 Explain how the changes that happen in the person's body cause the body temperature to fall. [2]

15–5 When a person's body temperature falls, they may begin to shiver. Explain how shivering can increase body temperature. [2]

Total: 10

Hormonal coordination in humans

Quick questions

p150 4.5.3.1 1 What is the endocrine system composed of?

p150 4.5.3.1 2 What is a 'hormone'?

p150 4.5.3.1 3 How are hormones transported?

p150 4.5.3.1 4 Where do hormones produce their effect?

p151 4.5.3.1 5 Which gland is known as the 'master gland'?

p152 4.5.3.2 6 Name the organ that monitors and controls blood glucose concentration.

p152 4.5.3.2 7 Name the hormone produced when blood glucose concentration is too high.

p152 4.5.3.2 8 What is excess glucose converted to for storage in humans?

9 Where do humans store excess glucose?

10 Describe the three main ways that water is lost from the body.

11 What is sweat composed of?

12 Name the organ that maintains the water balance of the body.

13 Name the main male reproductive hormone and state where it is produced.

14 Name the main female reproductive hormone and state where it is produced.

15 Which hormone is inhibited by oral contraceptives and what is the effect of inhibiting this hormone?

16 Name a barrier method of contraception.

17 What does a spermicide (or spermicidal agent) do?

H 18 Name the two hormones found in 'fertility drugs'.

H 19 Name the hormones produced by the adrenal glands and by the thyroid gland.

Exam-style questions

20 **Figure 6** shows some of the organs that make up the human endocrine system.

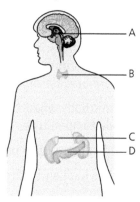

Figure 6

20–1 Name the type of organ that makes up the endocrine system. [1]

20–2 Identify organs A–D from the diagram. [4]

20–3 The endocrine system produces hormones.

Describe how hormones bring about an effect. [4]

20–4 Explain why the pituitary gland is described as the 'master gland'. [3]

H 20–5 The adrenal gland and the thyroid gland produce hormones.
Describe the roles of the hormones that each gland produces. [6]

H 20–6 Some people have hypothyroidism, which is an underactive thyroid. This mean that their thyroid gland does not produce enough of the hormone thyroxine. Suggest how the body of a person with hypothyroidism would try to respond to control the level of thyroxine in the blood. [3]

Total: 21

p151-3 `4.5.3.2`

21 **Figure 7** shows the changes in blood glucose concentration over time for a person who does not have diabetes.

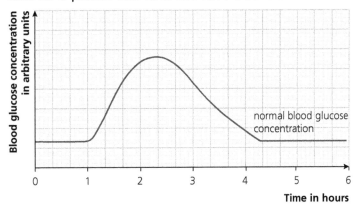

Figure 7

21-1 Suggest why the blood glucose concentration increased after 1 hour. *[1]*

21-2 Name the organ that would detect this increase in blood glucose concentration. *[1]*

21-3 Explain how the person's body responded to reduce the blood glucose concentration back to its normal level. *[2]*

H 21-4 Explain why the person's blood glucose concentration did not drop below the normal level. *[2]*

People with both type 1 and type 2 diabetes cannot control their blood glucose levels.

21-5 Describe how type 1 and type 2 diabetes are caused. *[2]*

21-6 Describe how type 1 and type 2 diabetes are treated. *[2]*

WS3.5
MS2c

21-7 Diabetes can be diagnosed using a glucose tolerance test. A person's blood glucose level is measured, then the person drinks a glucose drink and their blood glucose levels are measured at regular intervals.

Figure 8 shows the blood glucose level of a person who does not have diabetes after drinking a glucose drink.

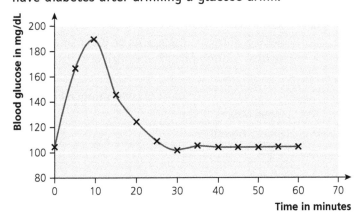

Figure 8

Describe **two** ways that the graph in **Figure 8** would be different in a person who **does** have diabetes after drinking a glucose drink. *[2]*

MS1b,1c **21–8** **Table 2** shows the number of people diagnosed with diabetes and the UK populations in 1998 and 2018.

	Year	
	1998	2018
Number of people in the UK diagnosed with diabetes	1.9×10^6	3.7×10^6
Population of the UK	5.8×10^7	6.5×10^7
Percentage of the population with diabetes	3.3	5.7

Table 2

Calculate the percentage of the UK population diagnosed with diabetes in 2018. *[3]*

MS1c **21–9** Calculate how many times greater the percentage of people with diabetes was in 2018 than in 1998. *[2]*

WS3.5 **21–10** Suggest **one** reason for the increase in the number of people with diabetes from 1998 to 2018. *[1]*

Total: 18

p155–8 4.5.3.3 **22** The kidneys control water balance in the body. Excess water is removed from the body via the kidneys in the urine.

22–1 Name **two** other substances that are removed from the body by the kidneys. *[2]*

22–2 Describe **two** other ways that water is lost from the body. *[2]*

22–3 Explain what could happen to our body cells if the blood was too dilute. *[2]*

H 22–4 A hormone called anti-diuretic hormone (ADH) is released by the pituitary gland when there is too little water in the blood. Explain how ADH can cause a change in the water content of the blood. *[3]*

H 22–5 The kidneys also controls nitrogen balance in the body by excreting a nitrogen-containing chemical produced when excess amino acids are broken down. Describe how the nitrogen-containing chemical is produced. *[3]*

WS3.5 **22–6** The kidneys filter the blood to produce filtrate, then use the filtrate to produce urine.

Table 3 shows the amount of glucose and protein in blood plasma, kidney filtrate and urine.

	Glucose concentration in g dm⁻³	Protein concentration in g dm⁻³
Blood plasma	0.9	75.0
Kidney filtrate	0.9	0.0
Urine	0.0	0.0

Table 3

Explain the glucose concentration of the kidney filtrate and urine. *[2]*

WS3.5 **22–7** Proteins in the blood can be large molecules. Suggest the reason for the protein concentration of the kidney filtrate and urine. *[1]*

WS1.4

22-8 Some people suffer from kidney failure and their kidneys cannot function properly. Kidney failure can be treated by organ transplant or by using kidney dialysis.

Figure 9 shows how a dialysis machine works.

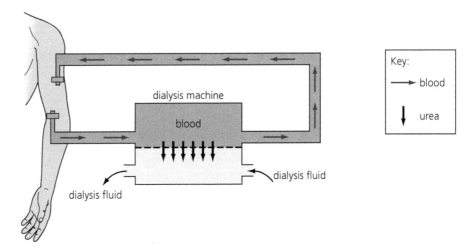

Figure 9

Describe how kidney dialysis works. Use information from the diagram and your own knowledge. [4]

WS1.3, 1.4
QWC

22-9 Suggest advantages and disadvantages of the two treatments for kidney failure. [6]

Total: 25

p158–60 4.5.3.4

23 Several hormones are involved in the menstrual cycle of a woman including follicle stimulating hormone (FSH), luteinising hormone (LH), oestrogen and progesterone.

Figure 10 shows the concentrations of hormones during the menstrual cycle.

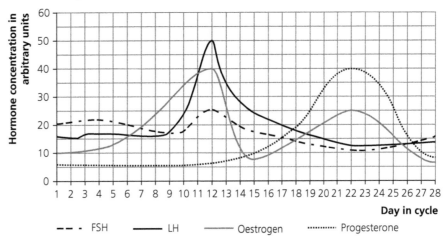

Figure 10

23-1 Name the organ that produces oestrogen. [1]

WS3.2
MS2a,2c

23–2 Calculate the rate of increase in the concentration of oestrogen between day 1 and day 12 in arbitrary units per hour. Give your answer to 3 significant figures. [2]

WS3.5

23–3 Identify the process that occurs on day 12 of the menstrual cycle. [1]

QWC (H) 23–4 Describe how FSH, LH, oestrogen and progesterone interact to control the menstrual cycle. [6]

23–5 Explain what happens when the concentrations of oestrogen and progesterone drop after day 22. [2]

Total: 12

p161–2 **4.5.3.5**

24 Contraception can be used to prevent unwanted pregnancy.

Table 4 gives information about different methods of hormonal and non-hormonal contraception.

Method of contraception	How it works	Percentage effectiveness with typical use
Oral contraceptive pill	Contains hormones and taken as a tablet for 21 days of the 28 day menstrual cycle.	91
Contraceptive implant	Contains slow release progesterone to inhibit the maturation and release of eggs for a number of months or years.	> 99
Condom	Barrier to prevent the sperm reaching an egg.	82
Intrauterine device (IUD)	A small plastic device inserted into the uterus where it prevents the implantation of an embryo.	> 99
Sterilisation	Sperm ducts or fallopian tubes are surgically cut so that eggs and sperm cannot meet.	> 99

Table 4

WS3.5

24–1 Identify the least reliable method of contraception from **Table 4**. [1]

24–2 Oral contraceptives contain hormones. Explain how the oral contraceptive pill works. [2]

WS1.4

24–3 Suggest **one** reason why the contraceptive implant is more effective than the oral contraceptive pill. [1]

24–4 Condoms can be used to prevent unwanted pregnancy. Give **one** other advantage of using condoms. [1]

WS1.3

24–5 Suggest why some people may have ethical issues with the intrauterine device (IUD). [1]

WS1.4

24–6 Suggest **one** disadvantage of sterilisation. [1]

Total: 7

p159–63 **2.5.3.4**

25 Some couples may experience infertility and the woman cannot become pregnant. The couple can have in vitro fertilisation (IVF) treatment which involves giving a woman hormone injections containing FSH and LH to stimulate the maturation of several eggs. Doctors collect the eggs from the woman in hospital.

25–1 State where the hormones FSH and LH are normally produced. [1]

(H) 25–2 Describe how doctors use the mature eggs in IVF treatment. [3]

WS1.4 (H) 25–3 Suggest why doctors only insert one or two embryos in the mother. [1]

WS1.4 **H** 25–4 Although fertility treatment gives a woman the chance to have a baby, there are disadvantages to the treatment.

Describe **two** disadvantages of IVF. [2]

Total: 7

Plant hormones

Quick questions

p170 4.5.4.1

1 What are the plant growth responses to light and gravity called?

p170 4.5.4.1

2 Which plant hormones affect plant growth and are involved in plant growth responses?

p171 4.5.4.1 **H**3 Which plant hormones are important in initiating seed germination?

p171 4.5.4.1 **H**4 Which plant hormone controls cell division and ripening of fruits?

p171 4.5.4.2

5 Give a use for each of these plant hormones: auxin, ethene, gibberellin.

Exam-style questions

p170–2 4.5.4.1 RP8

6 A student investigated plant growth responses using newly germinated seedlings.

The student:

- grew mustard seeds on a piece of cotton wool in a plastic dish for three days
- put the plastic dish of seedlings in a box with a hole cut in the side
- put a lamp next to the hole in the box
- left the seedlings in the box for five days.

Figure 11 shows the appearance of one of the seedlings at the start of the investigation and after five days in the box.

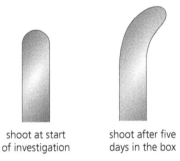

shoot at start of investigation

shoot after five days in the box

Figure 11

6–1 Describe the growth response of the shoot after five days. [2]

6–2 Explain why this growth response is useful for the shoot. [3]

6–3 Explain what causes the shoot to curve in this growth response. [3]

WS2.2 6–4 Suggest modifications to the method that would allow the student to investigate the effect of **light intensity** on the growth responses of mustard seeds. [2]

WS2.2 6–5 Suggest **two** control variables the student should use when investigating the effect of light intensity. [2]

Total: 12

p171–3 4.5.4.2

7 Plant hormones are used in agriculture and horticulture. Auxins are plant hormones that can be used as selective weed killers.

7–1 Give **two** other uses of auxins. [2]

WS1.3, 1.4

7–2 Selective weed killers kill some plants. Explain how the use of weed killers affects biodiversity. [2]

7–3 Ethene is a plant growth hormone used in the food industry. Explain why bananas are picked and transported when they are green, then stored in a room containing ethene gas. [2]

WS1.4

7–4 Gibberellins are plant hormones that have a number of uses in horticulture.

Suggest why a gardener soaks seeds in solution containing gibberellins before they are planted. [1]

WS1.4

7–5 Suggest why a gardener sprayed their tomato plants with a solution containing gibberellins. [1]

Total: 8

Homeostasis and response topic review

1 The endocrine system and nervous system both involve control systems that help to regulate the internal conditions of the body. All control systems have a coordination centre such as the brain or pancreas.

p132–3 4.5.1

1–1 Describe **two** other features that are common to all control systems. [2]

p132 4.5.2.1, 4.5.3.1

1–2 Describe **three** differences between a response by the endocrine system and a response by the nervous system. [3]

p141 4.5.2.3

1–3 A person is walking across a frozen lake when the ice below them cracks.

The person is looking into the distance when they hear the sudden crack of the ice breaking beneath them.

Describe the changes that happen inside the person's eye to allow them to focus on the ice near them. [3]

p135 4.5.2.1

1–4 Why does the person throw their arms out to the side as they begin to fall? [1]

p164 4.5.3.7

1–5 When the person begins to fall through the ice, adrenaline is released in response to the dangerous situation. Describe **two** effects that adrenaline has on the body. [2]

p144–5 4.5.2.4 QWC

1–6 When the person falls into the cold lake, the thermoregulatory centre in the brain detects a decrease in the person's core body temperature. Describe the responses to this decrease in temperature that will cause the temperature to return to normal. [6]

Total: 17

2   **Table 5** and **Table 6** show the volumes of water gained and lost by a person's body each day.

Source of water gain	Volume of water gained in cm³ per day
Chemical reactions	400
Drink	1500
Food	900

Table 5

Water output	Volume of water lost in cm³ per day
Exhaled air	500
Faeces	150
Sweat	700
Urine	X

Table 6

p155 4.5.3.3 WS3.2

2–1   Calculate the volume of water lost in urine in cm³ per day.   *[2]*

p155–6 4.5.3.3

2–2   The person runs a race, and loses a larger volume of water in sweat during the race.

Describe and explain the effect of increased sweating on the volume of urine produced.   *[3]*

p143 4.5.2.4

Ⓗ 2–3   Explain how sweating helps the runner to control their body temperature.   *[2]*

p143–4 4.5.2.4

2–4   Explain why the runner's face appears red during the race.   *[2]*

p152 4.5.3.2

Ⓗ 2–5   During the race, the runner's blood glucose level begins to fall. Describe how the runner's body will respond to the decrease in blood glucose levels.   *[3]*

p155 4.5.3.3

2–6   After the race, the runner drinks a sports drink that contains glucose and water. Suggest what else the sports drink should contain. Give a reason for your answer.   *[2]*

Total: 14

3   Hormones have an important role in human reproduction.

p158 4.5.3.4

3–1   Copy and complete **Table 7** to describe the main female and male reproductive hormones.   *[4]*

	Female	Male
Name of main reproductive hormone		
Site of production of hormone		

Table 7

p161–2 4.5.3.5 WS1.4

3–2 Hormones can be used as contraceptives to reduce fertility and prevent pregnancy.

Table 8 gives information about the progesterone-only oral contraceptive pill and the contraceptive implant.

	Progesterone-only pill	Contraceptive implant
How the contraceptive works	Prevents pregnancy by thickening the mucus in the cervix to stop sperm reaching an egg. Can also stop ovulation. The progestogen-only pill must be taken at the same time every day. If taken late it may not be effective.	A small flexible plastic rod is placed under the skin in your upper arm by a doctor or nurse. It releases the hormone progestogen into your bloodstream to prevent pregnancy and lasts for 3 years.
Advantages	More than 99% effective if taken correctly, but only about 91% effective with typical use. A pill is taken every day, with no break between packs of pills.	More than 99% effective. Can be left in place for 3 years. Useful for women who forget to take a pill at the same time every day. Can be taken out if there are any side effects. Can be removed at any time, and natural fertility will return quickly.
Disadvantages	May not work if you are sick or have severe diarrhoea. Some medicines may affect the progestogen-only pill's effectiveness. Periods may stop or become lighter, irregular or more frequent. May cause spotty skin and breast tenderness. Does not protect against sexually transmitted diseases.	Some bruising, tenderness or swelling around the implant when it is first inserted. Periods may become irregular, lighter, heavier or longer. Periods may stop. Some medicines can make the implant less effective. Does not protect against sexually transmitted diseases.

Table 8

Evaluate the use of the contraceptive implant compared to the progesterone-only pill. Use information from **Table 8**. [4]

p159–60&162 4.5.3.4, 4.5.3.6

Ⓗ 3–3 Hormones can be used in fertility drugs to increase the chance of pregnancy.

Describe how fertility drugs containing FSH and LH can be used to help a woman with infertility. [2]

p162–3 4.5.3.6 WS1.3

Ⓗ 3–4 In vitro fertilisation (IVF) can be used as a treatment for couples struggling to get pregnant naturally. This technique produces many embryos in a laboratory, but only one or two embryos are inserted back into the mother's uterus.

Suggest why some people are against the use of IVF. [1]

p170 4.5.4.1

3–5 Plants produce hormones to coordinate and control growth responses.

Describe how the growth responses of a newly germinated seedling can increase the chances of the seedling surviving. [4]

p171 4.5.4.1

3–6 Suggest **one** difference in the transport of plant hormones compared with human hormones. [1]

Total: 16

Inheritance, variation and evolution

Reproduction

Quick questions

p180	4.6.1.1	1 Which type of cell division forms non-identical cells?
p20&180	4.1.2.2, 4.6.1.1	2 Which type of cell division forms identical cells?
p180	4.6.1.1	3 What are the gametes in animals called?
p180	4.6.1.1	4 What are the gametes in plants called?
p178	4.6.1.1	5 What is 'asexual reproduction'?
p185	4.6.1.4	6 Describe the structure of DNA.
p183–4	4.6.1.4	7 Where is DNA contained in eukaryotic cells?
p184	4.6.1.4	8 What is a 'gene'?
p184	4.6.1.4	9 What does each gene code for?
p183	4.6.1.4	10 What is the 'genome' of an organism?
p185	4.6.1.5	11 What does a nucleotide consist of?
p185	4.6.1.5	12 What are the four bases that can be found in DNA?
p187	4.6.1.5	13 How many bases are needed to code for one amino acid?
p192	4.6.1.7	14 What is 'polydactyly'?
p192	4.6.1.8	15 How many pairs of chromosomes does an ordinary human body cell contain?

Exam-style questions

16 Organisms may reproduce asexually, sexually or both.

p181	4.6.1.1	16–1 In sexual reproduction, cells in the reproductive organs divide to form gametes. Describe the process of gamete formation. *[3]*
p179	4.6.1.1	16–2 Copy and complete the following sentences. Choose your answers from the options below. *[6]*

double	egg	fertilisation	meiosis	mitosis	pollen	single	sperm

The female gamete in humans is called the _____ and the male gamete is called _____.

Gametes are produced by _____ and contain a _____ set of chromosomes.

Gametes join together during a process called _____ and the new cell divides by _____.

p180	4.6.1.2	16–3 Human body cells have 46 chromosomes, but gametes only have 23 chromosomes. Explain why it is important for gametes to have 23 chromosomes. *[2]*

p21–2 4.6.1.2

16–4 A fertilised egg cell develops into a ball of cells called an embryo. Explain what happens to the fertilised egg cell as it develops into an embryo. *[2]*

p178–9 4.6.1.3

16–5 The parasite that causes malaria reproduces asexually in the human host, but sexually in the mosquito.

One advantage of asexual reproduction is that only one parent is needed. Describe **two** other advantages of asexual reproduction compared to sexual reproduction. *[2]*

p178–9 4.6.1.3

16–6 Describe another example of an organism that uses both asexual and sexual reproduction. *[2]*

p179 4.6.1.3

16–7 Explain why it is an advantage for a malaria parasite to reproduce sexually. *[4]*

Total: 21

17 DNA is a polymer. Polymers are made up of repeating units.

p185 4.6.1.5

17–1 Name the repeating units in DNA. *[1]*

p185 4.6.1.5

17–2 **Figure 1** shows the structure of a small section of DNA.

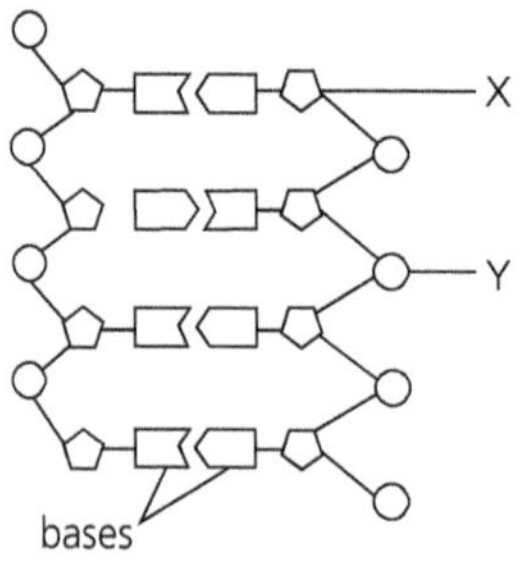

bases

Figure 1

Identify the parts labelled X and Y in **Figure 1**. *[2]*

p185 4.6.1.5

(H)**17–3** DNA contains four different bases. State the way the bases pair together. *[1]*

p186–7 4.6.1.5 QWC

(H)**17–4** The order of bases in DNA controls how proteins such as enzymes are made. Describe how the DNA base sequence is used to produce a specific enzyme. *[6]*

p187–8 4.6.1.5

(H)**17–5** Mutations occur continuously. A mutation occurs in a gene and causes a change in one amino acid of an enzyme.

Explain how a change in one amino acid could produce an enzyme that does not work properly. *[3]*

p184 4.6.1.5

(H)**17–6** Not all parts of DNA code for proteins. Explain the importance of these non-coding parts of DNA. *[2]*

p186 4.6.1.4 WS1.4

17–7 The genome is the entire genetic material of an organism. The whole human genome has now been studied and this has great importance for scientific understanding in the future, for example, in the search for genes linked to different types of disease.

Describe **two** other ways that understanding of the human genome is important. *[2]*

Total: 17

p188–
90 | 4.6.1.6

18 Genes are small sections of DNA that code for a specific protein. A gene often has two alleles.

18–1 Define the term 'allele'. [1]

Alleles can be expressed to give the phenotype of an organism.

18–2 Describe the alleles present for a recessive allele to be expressed in the phenotype. [1]

18–3 Most characteristics are a result of multiple genes interacting, but fur colour in mice is controlled by a single gene.

The allele for white fur is recessive and the allele for grey fur is dominant.

Use the following symbols to represent the alleles:

G = grey fur

g = white fur

A white male mouse breeds with a white female mouse. Give the reason why all of the offspring are white. [1]

WS3.5 | **18–4** Give the possible genotypes of a mouse with grey fur. [1]

WS1.2, 3.5 / MS2e | **18–5** A heterozygous male breeds with a homozygous recessive female. Copy and complete the Punnett square to determine the probability of them having white offspring. [3]

	G	g
g		
g		

WS3.5 | **18–6** Sickle cell disease (SCD) in humans is a blood disorder controlled by a single gene. SCD is caused by a recessive allele.

Use the following symbols to represent the alleles:

D = allele for normal health

d = allele for sickle cell disease

The family tree in **Figure 2** shows the inheritance of SCD.

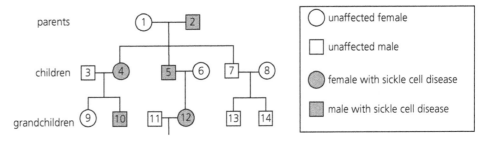

Figure 2

Individual 1 is a carrier of SCD. This means that individual 1 is heterozygous.

Explain why individual 7 is also a carrier of SCD. [2]

WS1.2, 3.5 / MS2e | **H 18–7** Individual 11 is **not** a carrier of SCD.

Draw a Punnett square to find the probability of individuals 11 and 12 having a child with SCD. [4]

Total: 13

19 The inheritance of certain alleles can cause inherited disorders.

Polydactyly is an inherited disorder caused by a dominant allele. People with polydactyly have extra fingers or toes.

A man with polydactyly and a woman without polydactyly have two children, one boy and one girl. One of the children has polydactyly and the other does not.

p189–92 | 4.6.1.6, 4.6.1.7 | WS3.5

19–1 Explain why the man must be heterozygous for polydactyly. *[3]*

p193 | 4.6.1.8

19–2 Give the sex chromosomes that would be found in a body cell of the boy and of the girl from question 19–1. *[2]*

p192–3 | 4.6.1.8 | MS1c,2e / WS1.2, 3.1

Ⓗ **19–3** The couple from question 19–1 decide to have another child. Use a genetic diagram to find the probability of the child being a girl. *[3]*

p188–92 | 4.6.1.6, 4.6.1.7 | WS1.2

19–4 Cystic fibrosis (CF) is an inherited disorder caused by a recessive allele.

Explain how two healthy carriers of the CF allele can have a child with CF. Use the symbols: **A** = allele for normal health; **a** = allele for CF. *[4]*

p188–92 | 4.6.1.6, 4.6.1.7 | MS1c,2e / WS1.2, 3.1

19–5 What is the probability of two healthy carriers of the CF allele having a child who is also a healthy carrier of the CF allele? Use your answer to question 19–4. *[1]*

p191–192 | 4.6.1.7

19–6 A couple who are both healthy carriers of the CF allele have embryo screening so that they can have a child without CF. Eggs from the mother are fertilised in a laboratory using sperm from the father. The fertilised egg will divide to form an embryo, then one cell is removed from the embryo and tested for the CF allele. Only embryos without the CF allele are put back into the mother's uterus.

Suggest **one** reason why people might be in favour of embryo screening. *[1]*

p191–2 | 4.6.1.7 | WS1.3

19–7 Suggest **one** reason why people might object to embryo screening. *[1]*

Total: 15

Variation and evolution

Quick questions

p198 | 4.6.2.1

1 What is 'variation'?

p200 | 4.6.2.1

2 How frequently do mutations occur?

p200 | 4.6.2.1

3 How frequently do mutations lead to a new phenotype?

p211 | 4.6.2.2

4 How long ago did simple life forms first develop?

p200–1 | 4.6.2.3

5 What is another term for artificial selection?

p183 | 4.6.2.5

6 Name **two** methods of cloning plants.

p183 | 4.6.2.5

7 What is 'tissue culture'?

p204–5 | 4.6.2.5

8 Name **two** methods of cloning animals.

Exam-style questions

9 Differences in the characteristics of individuals in a population is called variation.

p198 4.6.2.1

9–1 Describe the **two** causes of variation. *[2]*

There is usually extensive genetic variation within a population of a species. All variants happen due to changes in genes called mutations. Mutations occur continuously.

p200 4.6.2.1

9–2 Describe how likely it is that a mutation will impact the phenotype of an organism. *[2]*

p200 4.6.2.1

9–3 Explain how mutations can lead to rapid change in a species. *[3]*

p212–14 4.6.2.2

9–4 All species of living things have developed from simple life forms over millions of years by a process called evolution.

Define the term **evolution**. *[2]*

p212–14 4.6.2.2

9–5 Evolution can result in the formation of a new species.

Describe what it means when organisms are the same **species**. *[2]*

p212–14 4.6.2.2 WS1.2 QWC

9–6 The finch is a type of bird. Different species of finch live on different islands, but they evolved from a common ancestor.

Explain how two different species of finch could develop from a common ancestor. *[6]*

Total: 17

p200–1 4.6.2.3

10 Humans have been selectively breeding plants and animals for particular characteristics for thousands of years.

10–1 Give **three** useful characteristics that can be selected for. *[3]*

10–2 Selective breeding can be used to breed cattle that produce more meat or milk.

Describe how a farmer could selectively breed cattle that produce a large volume of milk. *[4]*

10–3 Selective breeding can lead to inbreeding. Give **one** reason why inbreeding is a disadvantage. *[1]*

Total: 8

p201–3 4.6.2.4

11 Genetic engineering involves modifying the genome of an organism by introducing a gene from another organism to give a desired characteristic.

Crops that have had their genes modified in this way are called genetically modified (GM) crops. Some plant crops have been genetically engineered to be resistant to diseases.

11–1 Give **two** other characteristics that have been introduced into GM plant crops. *[2]*

11–2 Bacterial cells have been genetically engineered to produce useful substances. Name **one** useful substance produced by genetically modified bacteria. *[1]*

WS1.3

11–3 Genetic modification of human cells is currently illegal, but modern medical research is exploring possible uses.

Suggest **one** possible benefit of genetic modification in humans. *[1]*

WS1.3

11–4 GM crops can have many useful characteristics, but some people object to them.

Describe **two** reasons why people may object to GM crops. [2]

QWC **H 11–5** Banana crops can be destroyed by a fungal disease called TR4. A gene from a nematode worm can cause resistance to the fungus that causes TR4. The nematode gene can be used to produce GM banana crops resistant to TR4.

Describe the main steps in the process of genetic engineering to produce banana crops resistant to TR4. [6]

Total: 12

12 Plants can be cloned using tissue culture or by taking cuttings.

p183 4.6.2.5 **12–1** Describe how plants are cloned using tissue culture. [2]

p183 4.6.2.5 **12–2** Give **one** reason why scientists might want to clone plants using tissue culture. [1]

p173&183 4.6.2.5 **12–3** Give **one** advantage of cloning plants by taking cuttings rather than using tissue culture. [1]

p205 4.6.2.5 **12–4** Animals can be cloned by either using embryo transplants or adult cell cloning.

Describe how animals are cloned using embryo transplants. [2]

p204–5 4.6.2.5 QWC **12–5** Adult cell cloning has been used to clone an endangered wild sheep called a mouflon.

Figure 3 shows how the scientists cloned a mouflon.

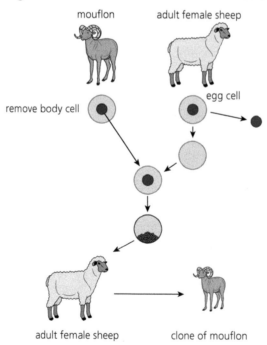

Figure 3

Describe how adult cell cloning could be used to clone a mouflon. Use your own knowledge and information from **Figure 3**. [6]

p204–5 4.6.2.5 WS1.3 **12–6** Evaluate the use of adult cell cloning to preserve endangered species like the mouflon. [4]

Total: 16

The development of understanding of genetics and evolution

Quick questions

p212   4.6.3.1   **1**   Name the theory proposed by Charles Darwin.

p213   4.6.3.1   **2**   What is the name of the book published by Charles Darwin with his theory of evolution, and when was it published?

p216   4.6.3.2   **3**   Which other scientist independently proposed the same theory of evolution as Charles Darwin?

p215   4.6.3.1   **4**   Which scientist developed a theory that changes that occur in an organism during its lifetime can be inherited?

p217   4.6.3.3   **5**   Which scientist studied inheritance of characteristics using breeding experiments in plants?

p219   4.6.3.5   **6**   What are fossils?

p222   4.6.3.6   **7**   What does it mean when a species is extinct?

p220   4.6.3.7   **8**   Why can bacteria evolve rapidly?

Exam-style questions

9   **Figure 4** shows a fossilised *Archaeopteryx*.

Figure 4

Fossils can tell us how much different organisms changed as life developed on Earth.

p220   4.6.3.5   WS1.3   **9–1**   Give **one** reason why scientists cannot be certain about how life began on Earth.   *[1]*

p219   4.6.3.5   **9–2**   Fossils may be formed as preserved traces of organisms, such as footprints, burrows and rootlet traces.

      Describe **two** other ways that fossils may be formed.   *[2]*

p222–3 | 4.6.3.6

9–3 The *Archaeopteryx* shown in **Figure 4** is extinct.

Give **three** possible causes of extinction. [3]

The fossil of the Archaeopteryx provides evidence for the origin of birds.

p212–13 | 4.6.3.1

9–4 Developing knowledge of geology and fossils helped Charles Darwin develop his theory of evolution by natural selection.

Describe **two** other sources of information that helped Darwin develop his theory. [2]

p214 | 4.6.3.1

9–5 Describe Darwin's theory of evolution by natural selection. [3]

p213 | 4.6.3.1 | WS1.1

9–6 In 1859, when Darwin published his theory of evolution in the book *On the Origin of Species*, there was a lot of controversy surrounding his revolutionary new ideas. One reason why the theory of evolution by natural selection was only gradually accepted is because it challenged the idea that God made all the animals and plants that live on Earth.

Describe **two** other reasons why the theory was only gradually accepted. [2]

p216–17 | 4.6.3.2

9–7 Alfred Russel Wallace also proposed the theory of evolution by natural selection and worked worldwide gathering evidence for evolutionary theory.

Give **one** other example of the work that Wallace is known for. [1]

p215–16 | 4.6.3.1

9–8 Jean-Baptiste Lamarck was another scientist who proposed a theory of evolution.

Describe Lamarck's theory of evolution. [2]

Total: 16

10 In the mid-nineteenth century, Gregor Mendel carried out breeding experiments on plants. He observed that the inheritance of each characteristic is determined by 'units' that are passed on unchanged to the offspring.

p218 | 4.6.1.4, 4.6.3.3

10–1 Name the 'units' that determine the characteristics of an organism. [1]

p218 | 4.6.3.3

10–2 Name the structures where these 'units' are located. [1]

p217–18 | 4.6.3.3 | WS1.1

10–3 Suggest **two** reasons why the importance of Mendel's discovery was not recognised until after his death. [2]

Total: 4

11 MRSA strains of bacteria are resistant to antibiotics.

p221 | 4.6.3.4 | WS1.3

11–1 Name the theory that is supported by the development of antibiotic resistance. [1]

p221 | 4.6.3.7

11–2 Explain how MRSA developed antibiotic resistance. [3]

p221 | 4.6.3.7

11–3 Give **one** reason why MRSA and other resistant strains of bacteria can spread. [1]

p221 4.6.3.7

11–4 One way to reduce the rate of development of antibiotic resistant strains of bacteria is to restrict the use of antibiotics in agriculture.

Describe **two** other actions that should be taken to reduce antibiotic resistance. [2]

p221 4.6.3.7

11–5 Scientists are trying to develop new antibiotics, but cannot do this quickly enough to keep up with the emergence of new resistant strains. Explain why. [2]

Total: 9

12 Charles Darwin visited the Galapagos Islands in the 1800s. He observed differences in the beaks of finches. The finches' beaks are different shapes so that they can eat different types of food.

Figure 5 shows three species of finch and the food the finches eat.

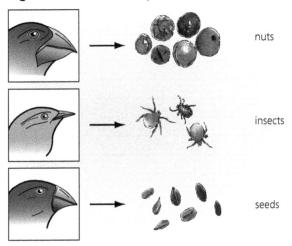

Figure 5

p214–16 4.6.3.1

12–1 Explain the evolution of the nut-eater beak shapes using:

- Darwin's theory [3]
- Lamarck's theory. [2]

p219–20 4.6.3.5

12–2 Fossils can give evidence for evolution.

Explain how a fossil of a finch could form. [3]

p219–20 4.6.3.4, 4.6.3.5

12–3 How can fossils give evidence for the evolution of finches' beaks? [1]

p222–3 4.6.3.6

12–4 Some species of finch are at risk of extinction due to human activity. Give **one** way that human activity could cause the extinction. [1]

p217 4.6.3.2

12–5 The insect-eating finches shown in **Figure 5** are similar to insect-eating finches on another island. How could scientists find out if the two groups of finches are the same species? [2]

p216–17 4.6.3.2

12–6 There are now 13 species of finches that evolved from a common ancestor. Explain how the different species of finch developed from a common ancestor. [7]

Total: 19

Classification

Quick questions

p228 4.6.4

1 Which scientist developed the traditional system of classifying living things into groups depending on their structure and characteristics?

p228 4.6.4

2 What is the name of the system of giving organisms two names based on their genus and species?

p230 4.6.4

3 Which scientist developed the 'three domain system' of classification?

Exam-style questions

p228–30 4.6.4

4 Carl Linnaeus developed a system of classifying living things into groups depending on their structure and characteristics.

Table 1 shows part of the classification of a tiger.

Kingdom	Animalia
1	Chordata
2	Mammalia
3	Carnivora
Family	Felidae
4	Panthera
Species	Tigris

Table 1

4–1 Give the names of the missing groups numbered 1–4 in **Table 1**. [4]

4–2 Give the binomial name of a tiger. Use information in **Table 1**. [1]

WS1.2

4–3 Evolutionary trees are used by scientists to show how organisms are related.

Figure 6 shows an evolutionary tree for the big cat group 'Panthera'.

Figure 6

Identify the big cat that is most closely related to a tiger based on the evolutionary tree in **Figure 6**. [1]

WS1.2

4–4 Identify the two big cats in **Figure 6** that share the most recent common ancestor. [1]

4–5 Due to evidence from developments in scientific techniques, there is now a 'three domain system' of classification developed by Carl Woese.

Describe the three domain system of classification. [3]

4–6 Suggest **one** scientific technique that may have provided evidence for the three domain system of classification. [1]

Total: 11

Inheritance, variation and evolution topic review

1   Cystic fibrosis (CF) is an inherited disorder.

A healthy couple have a child with CF.

p191-2 `4.6.1.6`

1-1 Explain how this shows that CF is caused by a recessive allele.   *[2]*

p189-90 `4.6.1.6` `WS1.2` Ⓗ

1-2 Draw a genetic diagram to show how the healthy couple could have a child with CF.   *[3]*

Use the symbols:

H = dominant allele and **h** = recessive allele.

`4.6.1.6` `MS2e`
`WS1.2`

1-3 The couple want to have another child. Give the probability of the couple having a child with CF. Use information from your genetic diagram.   *[1]*

p179 `4.6.1.1`

1-4 The couple decide to use embryo screening so that only an embryo without the CF allele will be put into the mother's uterus.

During embryo screening, gametes from the parents are joined by fertilisation and an embryo develops. One cell is removed from the embryo and screened for the CF allele.

Name the gametes produced by the parents.   *[1]*

p19 `4.6.1.2,`
`4.6.1.8`

1-5 Give the number of chromosomes that would be found in the cell from the embryo that is removed for screening. Explain your answer.   *[3]*

p192 `4.6.1.7`

1-6 Evaluate the use of embryo screening to have a child without CF.   *[4]*

p187-8 `4.6.1.5`

1-7 CF can be caused by the deletion of three bases from a gene called CFTR. The CFTR protein normally transports chloride ions across cell membranes. In people with CF, the CFTR protein cannot transport chloride ions in and out of cells properly.

Explain how the deletion of three bases can produce a protein that does not function properly.   *[3]*

Total: 17

2   Canthumeryx is an extinct ancestor of giraffes that lived 16 million years ago. **Figure 7** shows a Canthumeryx and a giraffe.

Figure 7

p222-3 `4.6.3.6`

2-1 Canthumeryx lived millions of years ago and are now extinct. Suggest **one** reason why Canthumeryx may have gone extinct.   *[1]*

p219-20 `4.6.4`

2-2 Suggest how scientists know what Canthumeryx looked like.   *[1]*

p214 `4.6.3.1` `WS1.1`

2-3 Charles Darwin and Jean-Baptiste Lamarck were both scientists in the nineteenth century who had different ideas about evolution.

Explain how the giraffe evolved to have a long neck using Darwin's theory of evolution by natural selection.   *[4]*

p215–16 4.6.3.1 WS1.1

2–4 Use Lamarck's theory of evolution to suggest how the giraffe evolved to have a long neck. *[2]*

p212–13 4.6.3.1 WS1.1

2–5 Other scientists in the nineteenth century also had theories of evolution.

Suggest **two** reasons why scientists in the nineteenth century were not sure which theory of evolution was correct. *[2]*

p229 4.6.3.5 MS4a

2–6 **Figure 8** shows the evolutionary tree for some organisms related to giraffes.

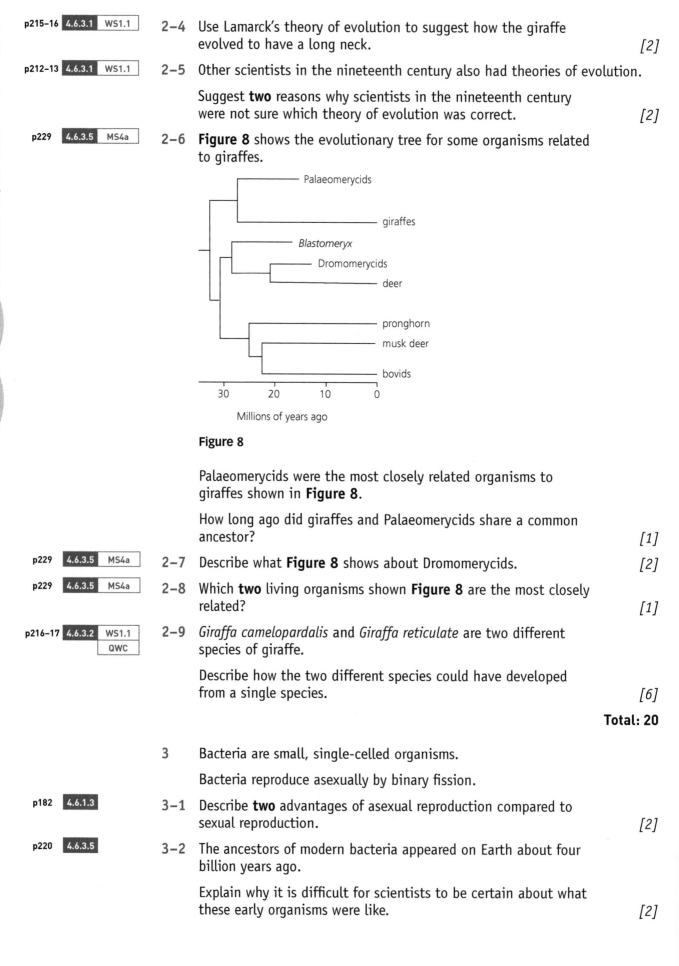

Figure 8

Palaeomerycids were the most closely related organisms to giraffes shown in **Figure 8**.

How long ago did giraffes and Palaeomerycids share a common ancestor? *[1]*

p229 4.6.3.5 MS4a

2–7 Describe what **Figure 8** shows about Dromomerycids. *[2]*

p229 4.6.3.5 MS4a

2–8 Which **two** living organisms shown **Figure 8** are the most closely related? *[1]*

p216–17 4.6.3.2 WS1.1 QWC

2–9 *Giraffa camelopardalis* and *Giraffa reticulate* are two different species of giraffe.

Describe how the two different species could have developed from a single species. *[6]*

Total: 20

3 Bacteria are small, single-celled organisms.

Bacteria reproduce asexually by binary fission.

p182 4.6.1.3

3–1 Describe **two** advantages of asexual reproduction compared to sexual reproduction. *[2]*

p220 4.6.3.5

3–2 The ancestors of modern bacteria appeared on Earth about four billion years ago.

Explain why it is difficult for scientists to be certain about what these early organisms were like. *[2]*

p230 `4.6.4`

3–3 Living things can be classified into groups. Carl Woese developed a 'three-domain system' of classification. One of the domains is 'bacteria'.

Name the **two** other domains in the 'three-domain system'. *[2]*

p3 `4.1.1.1`

3–4 **Figure 9** shows a bacterial cell.

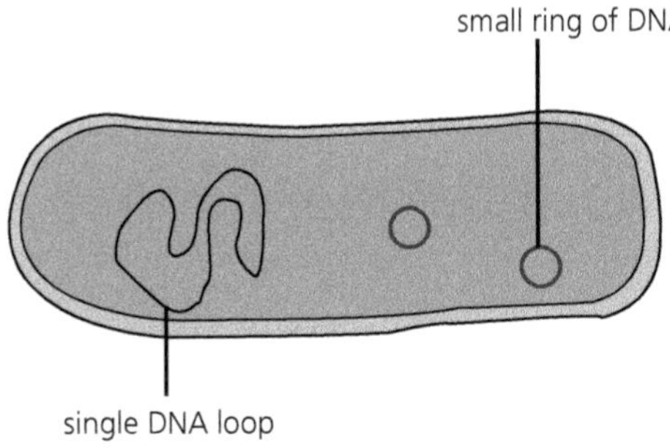

small ring of DNA

single DNA loop

Figure 9

Give the correct name of the small ring of DNA shown in **Figure 9**. *[1]*

p3 `4.1.1.1`

3–5 The bacterial cell has cytoplasm. What is the function of the cytoplasm? *[1]*

p3 `4.1.1.1`

3–6 Bacterial cells are prokaryotic. Give one piece of evidence from **Figure 9** that bacterial cells are prokaryotic. *[1]*

p301–2 `4.6.2.4`

Ⓗ**3–7** Scientists can use genetic engineering to produce human insulin.

Scientists can isolate the gene for human insulin.

Describe how scientists would use the isolated gene and the bacterial cell in **Figure 9** to produce insulin. *[4]*

p221 `4.6.3.7`

3–8 The small rings of DNA shown in **Figure 9** can carry genes for antibiotic resistance. One of the genes is for resistance to an antibiotic called methicillin.

Explain how the overuse of methicillin produced a population of bacteria that are resistant to the antibiotic. *[3]*

p221 `4.6.3.7`

3–9 Suggest **two** ways that doctors could help to reduce the rate of development of antibiotic resistant bacteria. *[2]*

Total: 18

p201–3 `4.6.2.4`

Ⓗ**4** Mosquitoes are insects that are vectors for malaria. Female mosquitoes bite humans to feed on blood.

A company has genetically engineered mosquitoes to produce genetically modified (GM) male mosquitoes. The GM male mosquitoes carry a gene from bacteria. The offspring of the GM males will die before they become adults.

4–1 Describe how the company used genetic engineering to produce GM mosquitoes. *[4]*

WS1.4
QWC

4–2 The GM male mosquitoes will be released into the wild where they will breed and pass on their genes.

Evaluate the release of the GM mosquitoes into the wild. *[6]*

Total: 10

Ecology

Adaptations, interdependence and competition

Quick questions

p236 4.7.1.1 1   What is a 'population'?

p236 4.7.1.1 2   What is a 'community?

p239 4.7.1.2 3   State what abiotic factors are and give **three** examples.

p241 4.7.1.3 4   State what biotic factors are and give **three** examples.

p251 4.7.1.1 5   Give the meaning of the word 'ecosystem', and give **two** examples of ecosystems.

p242–3 4.7.1.4 6   Explain what 'adaptations' are, and give examples of adaptations to extremely cold conditions.

p245 4.7.1.4 7   Explain what 'extremophiles' are and give an example.

Exam-style questions

p236–8 4.7.1.1 8   Organisms in a community often compete for resources.

8–1   Give **three** resources that plants compete with each other for.   [2]

8–2   Give **three** resources that animals compete with each other for.   [2]

8–3   Explain what the term 'interdependence' means.   [2]

8–4   Explain what makes a stable community.   [2]

WS1.2   8–5   **Figure 1** shows a simple grassland food web.

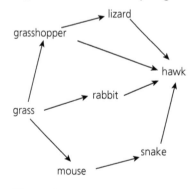

Figure 1

Explain the terms below with reference to the food web in **Figure 1**.

* community   [1]

* ecosystem   [1]

* population.   [1]

WS3.5 8–6 A disease kills all of the rabbits. Removing the rabbits from the food web might affect the mouse population. Suggest **one** reason why the mouse population might increase and **one** reason why the mouse population might decrease. [2]

Total: 13

9 Abiotic factors can affect communities.

p239–40 4.7.1.2 9–1 Give **three** examples of abiotic factors. [3]

p252–3 4.7.1.2 WS2.2 9–2 A student notices that there are less daisies growing in a shaded area under a tree than in an unshaded area.

A student makes the hypothesis:

'The number of daisies growing in a field is affected by light intensity.'

Describe an investigation to test this hypothesis. [6]

p111& 239 4.4.1.1, 4.4.1.2, 4.7.1.2 9–3 Explain why there might be fewer daisies in the shaded than unshaded area of the field. [2]

p264 4.7.1.2 MS4a 9–4 A scientist investigates the effect of sulfur dioxide concentration on the number of species of lichen that grow on trees.

The graph in **Figure 2** shows the scientist's results.

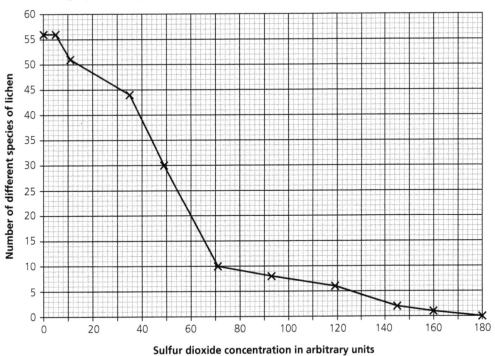

Figure 2

Use **Figure 2** to find the number of different species of lichen when the sulfur dioxide concentration is 60 arbitrary units. [1]

p264 4.7.1.2 WS3.2, 3.5 9–5 Describe the patterns in the data shown in **Figure 2**. [2]

p264 4.7.1.2 WS3.5 9–6 Give **one** conclusion that can be reached from the graph in **Figure 2**. [1]

Total: 15

p241–2 4.7.1.3

10 Biotic factors are living factors that can affect a community.

10–1 Give **three** examples of biotic factors. [3]

MS1c,2a
WS4.6

10–2 Two different species of squirrel are found in the UK, the red squirrel and the grey squirrel. The red squirrel is a native species whereas the grey squirrel was introduced from North America in the late nineteenth century.

The population of red squirrels has decreased from 3.5 million to 120 000.

Calculate the percentage decrease in the red squirrel population. Give your answer to 3 significant figures. [2]

WS1.2, 3.5

10–3 Table 1 gives information about grey and red squirrels.

	Grey squirrel	Red squirrel
Mass in g	540–660	270–340
Diet	Shares food sources with red squirrel and also eats seeds with high tannin content that red squirrels avoid.	Tree seeds, buds, flowers and shoots from trees, fungi, fruits.
Habitat	Both broad-leaved and coniferous forests.	Typically live in coniferous forests.
Population density per 10 000 m²	15 individuals	Up to three individuals
Disease risk	Carry the squirrelpox virus, but not affected by it.	Die within two weeks if infected with squirrelpox virus.

Table 1

Suggest **three** reasons why the population of red squirrels has decreased. Use information from the table. [3]

Total: 8

p244–5 4.7.1.4

11 Organisms are adapted to live in their natural environment.

Extremophiles live in extreme environments.

11–1 Give **three** examples of the features of extreme environments. [3]

11–2 Name **one** extremophile and state where it lives. [2]

WS3.5

11–3 The camel is an animal that lives in hot, dry conditions, such as a desert. Camels can feed on thorny desert plants and can live without water for months.

Figure 3 shows some of the adaptations that help the camel to survive.

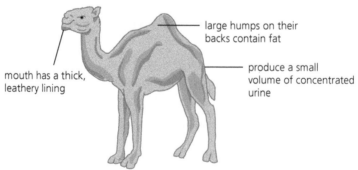

mouth has a thick, leathery lining

large humps on their backs contain fat

produce a small volume of concentrated urine

Figure 3

Suggest how each of the adaptations helps the camel to survive in the desert. [3]

11–4 Another adaptation of camels is that they avoid sitting in the Sun if possible.

State whether this is a structural, behavioural or functional adaptation. [1]

Total: 9

Organisation of an ecosystem

Quick questions

p288   4.7.2.1   1   Explain what is meant by the term 'trophic level', and state what the first trophic level in a food chain is.

p288   4.7.2.1   2   What is 'biomass' and what is the source of biomass on Earth?

p257   4.7.2.1   3   Define the term 'apex predator', and give an example of an apex predator.

p238–9   4.7.2.1   4   Describe what happens to the numbers of predators and prey in a stable community.

p258   4.7.2.2   5   Name the process that removes carbon from the atmosphere in the carbon cycle.

p258   4.7.2.2   6   Give **two** examples of processes that release carbon into the atmosphere in the carbon cycle.

p259   4.7.2.2   7   Give **two** examples of precipitation in the water cycle.

p257   4.7.2.3   8   Name the fuel produced by biogas generators.

Exam-style questions

p288   4.7.2.1   9   Feeding relationships within a community can be represented by food chains.

Look at this food chain:

grass → hare → lynx

9–1 Give the term used to describe the organism at the start of the food chain. [1]

9–2 Name the prey organism in the food chain. [1]

9–3 Name the secondary consumer in the food chain. [1]

9–4 Explain the importance of grass in the food chain. [2]

MS4a
WS3.2

9–5 The graph in **Figure 4** shows the changes in the populations of the lynx and the hare between 1890 and 1920.

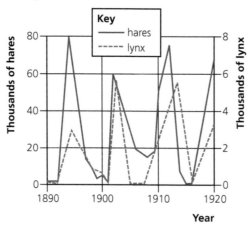

Figure 4

In 1894, there was a peak in both the lynx and hare populations.

Calculate how many times greater the hare population was than the lynx population.

Give your answer to 3 significant figures. *[3]*

MS4a
WS3.2

9–6 Describe the changes in the hare population between 1910 and 1915. *[2]*

MS4a
WS3.5

9–7 Suggest an explanation for the changes in the hare population between 1910 and 1915.

Use information from **Figure 4**. *[2]*

Total: 12

p253–5 4.7.2.1 RP9

10 A student uses a sampling technique to investigate the distribution of a plant called Ribwort plantain. The student uses this equipment:

- $0.25\,m^2$ quadrat

- 20 m tape measure.

QWC
AT3,6
WS2.2

10–1 Describe how the student would use the equipment to investigate the distribution of plantain across a footpath. *[6]*

WS2.3,
2.5

10–2 The student uses the equipment to estimate the number of plantain in a $400\,m^3$ field.

The student uses random sampling and counts the number of plantain in a large number of quadrats.

Give the reasons why the student uses **random sampling** and a **large** number of quadrats. *[2]*

WS3.3
MS2f

10–3 **Table 2** shows the student's results.

Quadrat number	1	2	3	4	5	6	7	8	9	10
Number of plantain	0	5	1	0	2	1	3	1	2	4

Table 2

Find the mode and median number of plantain per quadrat. *[2]*

WS3.3
MS2b,2f
10–4 Calculate the mean number of plantain per quadrat. *[1]*

WS3.3
MS2d
10–5 The area the student sampled was 20 m × 20 m.

Estimate the total number of plantain in this area. Use your calculated value for the mean number of plantain per quadrat. *[2]*

Total: 13

11 An ecosystem is the interaction of the abiotic and biotic parts of the environment.

Environmental changes affect the distribution of species in an ecosystem. Human activity can change the composition of gases in the atmosphere.

p262–4 4.7.2.4
11–1 Name **two** other environmental changes that can affect the distribution of species. *[2]*

p262–4 4.7.2.4
11–2 Human activity can change environmental conditions. Give **two** other reasons for changes in environmental conditions. *[2]*

p259–60 4.7.2.2 WS1.2
11–3 Many different materials cycle through the abiotic and biotic components of an ecosystem.

The water cycle provides fresh water for plants and animals on land before draining into the seas.

Figure 5 shows the water cycle.

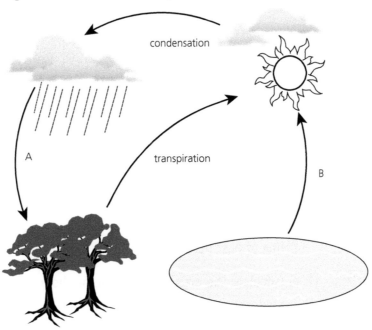

Figure 5

Name the processes labelled A and B on **Figure 5**. *[2]*

p72 4.2.3.2
11–4 One of the processes shown on **Figure 5** is transpiration. Explain what transpiration is. *[2]*

p258–9 4.7.2.2 QWC
11–5 The carbon cycle is important in recycling carbon for living organisms. Explain how the carbon in dead plants can become part of a living plant. *[6]*

Total: 14

12 Gardeners and farmers try to provide optimum conditions for the rapid decay of waste biological material to produce compost.

Compost heaps are regularly turned over to provide more oxygen for rapid decay.

12–1 Give **two** other conditions that will cause rapid decay. [2]

12–2 Explain what gardeners and farmers use compost for. [2]

12–3 Waste biological material can be put into a biogas generator to produce biogas.

Explain what happens to the waste material in a biogas generator. [2]

RP10
MS4a,4c
WS3.2

12–4 As milk decays, microorganisms break down lactose in milk to lactic acid, and the pH of the milk decreases.

A student carries out an experiment to investigate the effect of temperature on the rate of decay of fresh milk.

The student:

- pours 100 cm^3 of fresh milk into a sterile conical flask
- measures the pH of the milk using a pH meter
- puts the flask in a water bath at 20 °C
- measures the pH of the milk every day for the next five days.

Table 3 shows the student's results.

Time in days	pH of milk
0	6.7
1	6.7
2	6.6
3	6.3
4	5.5
5	4.8

Table 3

Plot a graph of the student's results.

You should choose suitable scales, label the axes and draw a line of best fit. [5]

RP10
MS1b

12–5 Calculate the rate of pH change per hour between days 0 and 5.

Give your answer in standard form. [3]

RP10
WS3.5

12–6 The student repeated the investigation at 40 °C and at 60 °C.

Explain why the rate of pH change would be:

- faster at 40 °C [2]
- slower at 60 °C. [2]

Total: 18

Biodiversity and the effect of human interaction on ecosystems

Quick questions

p270 | 4.7.3.1
1   What is 'biodiversity'?

p275 | 4.7.3.2
2   Which gases can cause acid rain?

p277 | 4.7.3.3
3   What is 'peat' and why have peat bogs been destroyed?

p270 | 4.7.3.4
4   What is 'deforestation' and why does it occur?

p279 | 4.7.3.5
5   What is 'global warming'?

p279–80 | 4.7.3.5
6   What are the main greenhouse gases?

Exam-style questions

7   The human population is using more resources and causing more pollution.

p271–5 | 4.7.3.2
7–1   Copy and complete **Table 4** to give **one** way that each of air, land and water can be polluted by humans.   [3]

Area polluted	Source of pollution
Air	
Land	
Water	

Table 4

p271 | 4.7.3.2
7–2   Give **two** reasons why humans are using more resources and producing more waste.   [2]

p271 | 4.7.3.2
7–3   Biodiversity is the variety of all the different species of organisms within an ecosystem.

Describe the effect that pollution has on biodiversity.   [2]

p270 | 4.7.3.1
7–4   Explain why a greater biodiversity results in a more stable ecosystem.   [2]

p276–7 | 4.7.3.3
7–5   Farming has reduced the amount of land available for other plants and animals.

Give **two** other human activities that have reduced land availability.   [2]

p277 | 4.7.3.3 | QWC
7–6   In the past, farmers used peat to produce compost that improves the quality of soil, as peat was cheap and readily available.

Explain the advantages and disadvantages of using peat in this way.   [6]

Total: 17

8   Human activity can affect ecosystems.

Large scale deforestation has occurred in tropical areas. In Brazil the percentage of land covered by forest decreased from 65.41% in 1990 to 59.05% in 2015.

The land area of Brazil is 8.5 million km^2.

pXX | 4.7.3.4 | MS1c
8–1   Calculate the area of forest lost in Brazil between 1990 and 2015.   [2]

p278   4.7.3.4

8–2   Give **two** reasons why large scale deforestation has occurred in tropical areas.   [2]

p270& 278   4.7.3.4

8–3   Deforestation can increase the greenhouse effect.

Describe **two** other negative effects that deforestation can have on ecosystems.   [2]

The greenhouse effect can lead to global warming.

p279   4.7.3.5

8–4   Describe **two** possible effects of global warming.   [2]

p277–80   4.7.3.3–5

8–5   Explain how human activities, other than deforestation, can increase the proportion of greenhouse gases in the atmosphere.   [3]

p279–80   4.7.3.5

8–6   Explain how increasing the proportion of greenhouse gases in the atmosphere can lead to global warming.   [3]

p280–2   4.7.3.6

8–7   Reducing deforestation and carbon dioxide emissions is one way of reducing the negative effects that humans can have on ecosystems and biodiversity.

Describe **three** other ways of reducing the negative effects of humans on ecosystems and biodiversity.   [3]

Total: 17

Trophic levels in an ecosystem

Quick questions

p288   4.7.4.1

1   Explain what a pyramid of biomass is.

p288   4.7.4.1

2   What are primary consumers?

p288   4.7.4.1

3   What are carnivores?

p288   4.7.4.1

4   What are apex predators and where are they found in a food chain?

Exam-style questions

5   Food chains can be used to represent feeding relationships.

Look at this marine food chain:

phytoplankton → zooplankton → sardine → tuna

Table 5 shows the biomass of each organism in the food chain.

Organism	Biomass in $g\,m^{-2}$
Phytoplankton	840
Zooplankton	480
Sardine	60
Tuna	12

Table 5

p257   4.7.4.1

5–1   Give an example of an organism from the food chain that can be described using each of the scientific terms below.   [3]

Apex predator	Producer	Secondary consumer

p288-9 | 4.7.4.2 | WS1.2 | MS2c

5–2 Draw a pyramid of biomass for the food chain.

You need to:

- use a suitable scale
- label the *x*-axis
- label each trophic level. [4]

4.7.4.3 | MS1c

5–3 Calculate the percentage of biomass lost between the phytoplankton and the tuna.

Give your answer to 2 significant figures. [3]

p290 | 4.7.4.3

5–4 Describe **two** ways that biomass can be lost between trophic levels. [2]

p257 | 4.7.4.1

5–5 Dead plant and animal matter is broken down by decomposers.

Describe how decomposers break down dead matter. [3]

p290 | 4.7.4.3

5–6 Humans can eat both sardines and tuna.

Explain why a greater proportion of biomass is passed to humans if they eat sardines rather than tuna. [3]

pXX | 4.4.1.1, 4.7.4.3 | WS3.5

5–7 Plants and algae only transfer about 1% of the energy from light to glucose by photosynthesis.

Suggest **one** reason why only 1% of light energy is transferred by photosynthesis. [1]

Total: 19

Food production

Quick questions

p296 | 4.7.5.1

1 What is meant by 'food security'?

p298 | 4.7.5.2

2 Why are some animals fed high protein foods?

p298 | 4.7.5.2

3 What is meant by 'intensive farming'?

p301 | 4.7.5.4

4 What is 'mycoprotein'?

p301 | 4.7.5.4

5 Which hormone to treat diabetes is purified and harvested from genetically modified bacteria?

Exam-style questions

6 Food security is having enough food to feed the population.

The increasing birth rate is one biological factor that is threatening food security in some countries.

p296-7 | 4.7.5.1

6–1 Describe **three** other factors that are threatening food security. [3]

p298 | 4.7.5.2

6–2 Sustainable farming methods must be found to feed all people on Earth.

Animals can be intensively farmed to increase the efficiency of food production.

Explain how intensively farming animals can increase the efficiency of food production. [4]

p298 | 4.7.5.2 | WS1.3, 1.4

6–3 Suggest **two** disadvantages of intensively farming animals. [2]

Total: 9

p300 4.7.5.3

7 Fish stocks in the oceans are declining. It is important to maintain fish stocks.

Figure 6 shows a new type of fishing net.

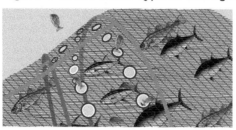

Figure 6

WS1.4

7–1 Explain how the new type of fishing net can help to maintain fish stocks. [2]

7–2 Fish stocks can be maintained by controlling net size.

Give **one** other way of maintaining fish stocks. [1]

Total: 3

p301 4.7.5.4

8 Modern biotechnology techniques enable large quantities of microorganisms to be cultured for food.

Mycoprotein is a protein-rich food grown in a fermenter, then harvested and purified.

8–1 Name the type of microorganism that produces mycoprotein. [1]

8–2 The fermenter is sterilised before adding the microorganisms to make mycoprotein.

Explain why. [2]

8–3 Glucose syrup is added to the fermenter. Give the reason why. [1]

8–4 Air is bubbled into the fermenter. Give **two** reasons why. [2]

8–5 Suggest **one** reason why a cooling jacket is needed. [1]

QWC 8–6 Describe the advantages of eating mycoprotein rather than eating animal products. [6]

Total: 13

Ecology topic review

1 Marram grass is a plant that grows on sand dunes next to the sea.

The distribution of marram grass is affected by abiotic (non-living) factors.

p239–40 4.7.1.2

1–1 Give **two** abiotic factors that could affect the distribution of marram grass. [2]

p242 4.7.1.4

1–2 Marram grass has adaptations to help it survive in the dry conditions of the sand dunes.

The adaptations include:

• leaves covered with a thick, waxy cuticle

• very long roots.

Explain how these adaptations mean that marram grass can survive in dry, sandy conditions. [2]

p253& 255 | 4.7.2.1 | RP9
WS2.5
MS2d

1–3 A student investigates the distribution of the plants marram grass and rest harrow on sand dunes.

The graph in **Figure 7** shows the student's results.

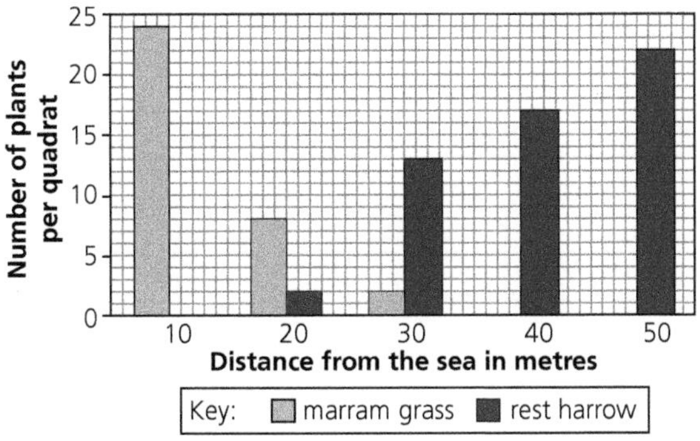

Figure 7

Describe how the student used a quadrat to collect the results shown in the graph. [3]

WS3.2, 3.5 | MS4a

1–4 Describe the pattern shown in the results for marram grass. [2]

p242 | 4.7.1.1, 4.7.1.3 | WS3.5

1–5 Suggest an explanation for the change in the distribution of marram grass. [2]

Total: 11

2 Human activities can have many effects on ecosystems.

Raised peat bogs cover 7000 km^2 of Britain.

Over the last 100 years, 94% of raised peat bogs have been lost in Britain.

WS3.3

2–1 Calculate the rate of loss of raised peat bogs in Britain over the last 100 years in km^2 per year.

Give your answer to the nearest whole number. [2]

p270& 277 | 4.7.3.3

2–2 Many peat bogs have been drained and are now used for agriculture.

Describe and explain how draining peat bogs affects biodiversity. [2]

p270 | 4.7.3.1

2–3 Explain why it is important to maintain biodiversity in an ecosystem. [2]

p277& 281 | 4.7.3.6

2–4 Suggest one way that humans could reduce the negative effects they have on peat bogs. [1]

p202 | 4.6.2.4, 4.7.5.4

2–5 Crops can be genetically modified to improve their nutritional content; for example, golden rice has been genetically modified to prevent vitamin A deficiency.

Suggest **one** other advantage of growing genetically modified crops. [1]

p202 | 4.6.2.4, 4.7.5.4 | WS1.3

2–6 Suggest **one** reason why people may be against genetically modified crops. [1]

Total: 9

3   Deforestation is the clearing of forests on a large scale.

Table 6 gives the area of deforestation in the Brazilian Amazon between 1985 and 2015.

Year	Area of deforestation in 1000 × km²
1985	21.1
1990	13.7
1995	29.1
2000	18.2
2005	19.0
2010	7.0
2015	6.1

Table 6

MS2c

3–1   Plot a bar chart of the data given in **Table 6**.   [5]

WS3.2, 4.3

3–2   Calculate the total area of forest cleared in the Brazilian Amazon from 1985 to 2015.   [2]

p278   4.7.3.4

3–3   Give **two** reasons why deforestation occurs.   [2]

p278   4.7.2.2, 4.7.3.5   QWC

3–4   Describe how deforestation can change the composition of gases in the atmosphere.   [6]

p279   4.7.2.4, 4.7.3.5

3–5   Changing the composition of gases in the atmosphere can lead to global warming.

Describe **two** effects of an increase in the Earth's temperature.   [2]

Total: 17

4   Overfishing removes a species of fish faster than it can be replaced.

p300   4.7.5.3

4–1 State **two** regulations that would help to conserve fish stocks at
     sustainable levels.   [2]

p288–9   4.7.4.3

4–2 Salmon for human consumption are farmed in the UK in open sea cages.

Figure 8 shows an open sea cage.

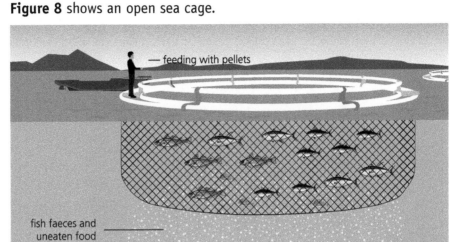

Figure 8

The salmon are fed pellets containing vegetable protein and
smaller fish such as anchovy. Anchovy eat plankton.

This is the food chain:

plankton → anchovy → salmon → human

Draw and label the pyramid of biomass for this food chain.   [2]

p290   4.7.4.3

4–3 Give **two** reasons why it would be more efficient if humans ate
     anchovies rather than feeding them to salmon.   [2]

p298   4.7.5.2

4–4 Explain why farmed salmon in sea cages grow faster than wild
     salmon in the sea.   [4]

p298   4.7.3.2   WS1.3

4–5 Some people are concerned about the environmental impact of
     farming salmon in sea cages.

Suggest **one** reason why people might be concerned.   [1]

Total: 11

Practice exam papers

Paper 1

1   **Figure 1** shows a human cheek cell as seen using a light microscope.

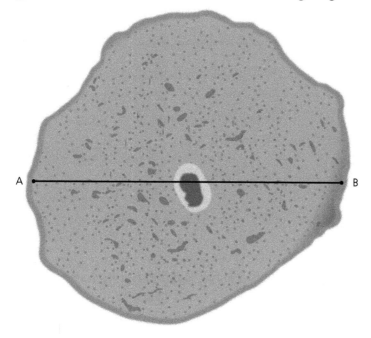

Figure 1

1–1   Draw a scientific drawing of the cheek cell.

Label the nucleus and cell membrane on your drawing.   *[2 marks]*

1–2   The real diameter of the cheek cell in **Figure 1** is 60 µm between points **A** and **B**.

Calculate the magnification of the cheek cell.   *[3 marks]*

1–3   Name **one** structure found in a root hair cell that is not found in a cheek cell.   *[1 mark]*

The cheek cell contains ribosomes and a nucleus.

1–4   Give the function of ribosomes.   *[1 mark]*

1–5   Name the group of living things that do not have their genetic material enclosed
in a nucleus.   *[1 mark]*

1–6   Give **two** reasons why ribosomes can be seen using an electron microscope, but not
using a light microscope.   *[2 marks]*

Total: 10

2 Measles is an infectious disease that can be fatal.

2–1 Name the type of pathogen that causes measles. *[1 mark]*

2–2 Young children are given the MMR vaccine to protect them against measles, mumps and rubella.

Explain how vaccination can prevent illnesses such as measles. *[6 marks]*

Read this information:

In 1998, Andrew Wakefield claimed there was a link between the MMR vaccine and autism.

Autism affects brain function and can cause difficulties with communication and behaviour.

Wakefield based his claim on the cases of 12 children. The parents of eight of the children said their child started showing signs of behaviour change after receiving the MMR vaccination.

Wakefield's work has been completely discredited.

In a recent study, researchers followed 657 461 Danish children until they were on average eight years old. 95% of children in the study were vaccinated against MMR.

Around 1 in 100 of the children in the Danish study developed autism, but there was no difference in the rates of autism between those who had been vaccinated and those who had not.

2–3 Suggest **two** reason why Wakefield's claim that the MMR vaccine causes autism is not valid. *[2 marks]*

2–4 Calculate the approximate number of children in the Danish study who developed autism. *[1 mark]*

2–5 The World Health Organisation recommends that at least 95% of children are vaccinated against MMR.

In 2018, the vaccination rate for MMR in England was 91.2%

Suggest **one** reason why parents may choose not to vaccinate their children. *[1 mark]*

Total: 11

3 Photosynthesis produces glucose and oxygen.

3–1 Give **two** ways that plants use the glucose produced in photosynthesis. *[2 marks]*

3–2 The rate of photosynthesis can be affected by factors including temperature and light intensity.

Sketch a graph to show the effect of temperature on the rate of photosynthesis.

Label the axes 'Temperature' and 'Rate of photosynthesis'.

You do not need to include scales on the axes. *[1 mark]*

3–3 A student investigates the effect of light intensity on the rate of photosynthesis by recording the number of bubbles of gas produced by pondweed at different light intensities.

Figure 2 shows the apparatus the student uses.

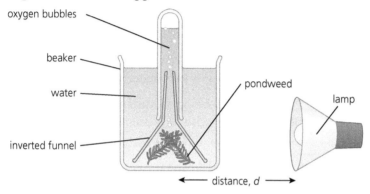

Figure 2

The student:
- sets up the apparatus as shown in **Figure 2**
- puts the lamp at a distance of 30 cm from the pondweed
- turns the lamp on and waits for 2 minutes
- counts the number of bubbles of gas produced in 10 minutes
- repeats with the lamp at different distances from the pondweed.

Identify the dependent variable in the student's investigation. *[1 mark]*

3–4 The teacher suggests putting a thermometer in the beaker of water.

Suggest the reason why this would make the student's results more valid. *[1 mark]*

3–5 The student compares their results to another student to check they are reproducible.

Suggest **one** variable the students would need to control so they can compare results. *[1 mark]*

Table 1 shows the student's results.

Distance, d, between the lamp and the pondweed in m	Light intensity in arbitrary units	Number of bubbles produced in 10 minutes
0.3	11	52
0.4	6	48
0.5	4	37
0.6	3	25
0.7	X	13

Table 1

Light intensity can be calculated using the inverse square law:

$$\text{light intensity} = \frac{1}{d^2}$$

where d is the distance between the lamp and the pondweed.

3–6 Calculate the light intensity when the distance between the lamp and the pondweed is 0.7 m. *[1 mark]*

3–7 Plot a graph to show the effect of light intensity on the number of bubbles produced.

Use data from **Table 1** and your answer to Question 3–6.

You should:

- use suitable scales
- label the axes
- draw a line of best fit. *[4 marks]*

3–8 Predict the number of bubbles that would be produced when the distance between the lamp and the pondweed is 0.2 m. *[1 mark]*

Total: 12

4 Substances are transported around plants.

4–1 Describe how food molecules are transported around plants. *[2 marks]*

4–2 Plants lose water through their leaves. Describe how plants control the amount of water lost from their leaves. *[2 marks]*

4–3 A student investigates the loss of water from the leaves of two plants, plant **A** and plant **B**.

The student puts the two plants in different conditions.

Figure 3 shows the student's results.

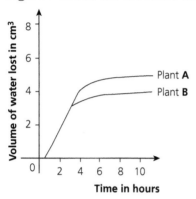

Figure 3

Calculate the rate of water loss from plant **A** in cm³ per hour.

Give your answer to 2 significant figures. *[3 marks]*

4–4 Plant **A** and plant **B** are both the same species of plant.

Suggest **two** environmental conditions that would explain the difference in water loss between plants **A** and **B**, apart from the different concentration of carbon dioxide. *[2 marks]*

4–5 Suggest **one** reason why different plant species might have different rates of water loss. *[1 mark]*

Total: 10

5 In coronary heart disease (CHD) layers of fatty material can build up inside the blood vessels supplying the heart muscle.

Figure 4 shows a blood vessel from a person with CHD.

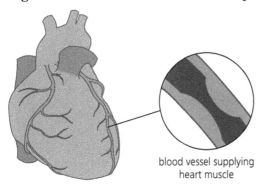

blood vessel supplying
heart muscle

Figure 4

5–1 Name the blood vessels that supply the heart muscle with blood. *[1 mark]*

5–2 Explain why CHD can prevent the heart from functioning. *[4 marks]*

CHD can be treated using statins or stents. **Table 2** gives information about each treatment.

Statins	Stents
Statins are given as tablets that are taken once a day for life. Statins can interact with other drugs including some antibiotics. Common side effects of taking statins include nose bleeds, sore throat, nausea and digestive problems. Rare side effects include liver and kidney damage. Statins can also cause muscle swelling and damage. A scientific study into the effectiveness of statins found that 232 out of 3794 men aged 60–69 who take statins had a serious cardiac event such as a heart attack or stroke during the study.	A stent is a small hollow tube. To insert a stent a doctor makes an incision (cut) in a blood vessel in the patient's leg and the stent is moved towards the heart using a wire inside the artery. Risks of inserting stents include damage to blood vessels in the leg or heart, and there is up to a 2% chance of having a heart attack during the procedure. After having a stent, the patient will need to take blood-thinning medication for up to one year and take aspirin every morning for life to reduce the risk of blood clots forming around the stent. A scientific study into the effectiveness of stents found that 224 out of 2493 men aged 60–69 had a serious cardiac event during the study.

Table 2

5–3 Give **one** conclusion about the effectiveness of the two treatments in preventing a serious cardiac event.

Use the information provided.

Include calculations to support your answer. *[3 marks]*

5–4 Evaluate the use of statins and stents to treat CHD.

Use information from **Table 2** and your own knowledge. *[6 marks]*

Total: 14

6 Multicellular organisms have specialised surfaces for exchange of materials.

6–1 Give the reason why a single-celled organism does not need a specialised exchange surface. *[1 mark]*

6–2 Proteins are broken down to amino acids.

Describe how you could test a solution to show that it contains protein. *[2 marks]*

6–3 Amino acids are absorbed into the blood.

Name the part of the blood that transports amino acids. *[1 mark]*

6–4 Amino acids enter the blood by diffusion and active transport.

Describe how the small intestine is adapted for the rapid absorption of amino acids. *[5 marks]*

6–5 Cholera is a bacterial disease that affects the cells lining the small intestine.

Cholera causes an increase in transport of chloride ions out of the cells and into the lumen of the intestine. Water moves into the lumen causing diarrhoea.

Explain how cholera causes water to move out of cells. *[3 marks]*

Total: 12

7 **Figure 5** shows the human breathing system.

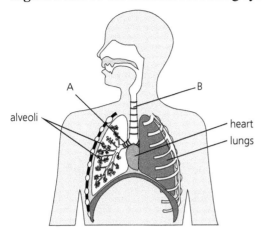

Figure 5

7–1 Name parts A and B shown on **Figure 5**. *[2 marks]*

7–2 Each lung in a human contains about 350 million alveoli.

The lungs provide a total surface area of $140\,m^2$.

Calculate the area of one alveolus. Give your answer in standard form. *[2 marks]*

7–3 **Figure 6** shows an alveolus from a healthy person and an alveolus from a person with damaged lungs.

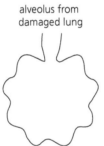

Figure 6

Explain why gas exchange is less efficient in the person with damaged lungs. *[2 marks]*

7–4 As a person climbs a mountain, the volume of oxygen in each breath decreases.

Acute mountain sickness (AMS) can occur at heights above 2500 m due to the decrease in oxygen.

Symptoms of AMS include headache, nausea (feeling sick) and extreme tiredness.

Explain why people with AMS may feel tired. *[4 marks]*

7–5 The extremely low temperatures on some mountains can freeze the skin and cause frost bite.

Severe frost bite can cause tissue death and may require a transplant of healthy skin.

Suggest how cells from human embryos could be used to produce healthy skin to treat frost bite. *[2 marks]*

Total: 12

8 Lipase is an enzyme that breaks down lipids (fats).

8–1 Name the products of lipid digestion. *[2 marks]*

8–2 The 'lock and key theory' is a model to explain enzyme action.

Describe how lipase breaks down lipids using the 'lock and key theory' of enzyme action. *[2 marks]*

8–3 Bile is a chemical that helps lipase to digest lipids.

Bile is made in the liver and stored in the gall bladder.

Some people have a blocked bile duct and bile cannot enter the small intestine.

One of the symptoms of a blocked bile duct is weight loss.

Explain how a blocked bile duct may cause weight loss. *[4 marks]*

Total: 8

9 Cancer is a disease that leads to uncontrolled growth and division of cells producing tumours.

9–1 Give **two** differences between benign and malignant tumours. *[2 marks]*

9–2 A scientist develops a new drug to treat cancer. The drug is tested in a clinical trial.

The patients in the clinical trial are divided into two groups:

- Group **A** are given the drug directly into the blood stream through a drip.

- Group **B** are given a placebo.

- Group **A** and Group **B** are matched for as many factors as possible to ensure valid results.

Suggest **one** factor that Group **A** and Group **B** should be matched for. *[1 mark]*

9–3 Suggest a suitable placebo for the clinical trial. *[2 marks]*

9–4 Some cancers can be treated with drugs attached to monoclonal antibodies.

The monoclonal antibodies deliver the drug to cancer cells.

To make monoclonal antibodies a scientist first isolates proteins from the surface of cancer cells and injects the proteins into a mouse.

Explain how the scientist uses cells from the mouse to make monoclonal antibodies. *[3 marks]*

9–5 Monoclonal antibodies can be bound to toxic chemicals that kill cancer cells, or stop them from dividing and growing.

A common side effect of some monoclonal antibodies is an allergic reaction to the drug. The symptoms of this reaction may include fever, an itchy rash and breathlessness.

Evaluate the use of monoclonal antibodies to treat cancer.

Use the information provided and your own knowledge. *[3 marks]*

Total: 11

Total for Paper 1: 100

Paper 2

1   **Figure 1** shows a food chain (not drawn to scale).

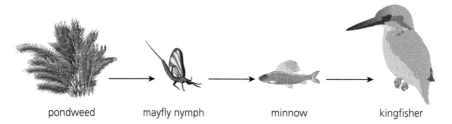

pondweed   mayfly nymph   minnow   kingfisher

Figure 1

1–1 Name the primary consumer in the food chain.   *[1 mark]*

1–2 Draw a pyramid of biomass for the food chain shown in **Figure 1**.

Label each trophic level.   *[2 marks]*

1–3 Give **one** way that biomass is lost between trophic levels.   *[1 mark]*

A student investigates the number of mayfly nymphs in a pond.

The student:

- collects $100 \, cm^3$ of pond water

- counts the number of mayfly nymphs

- repeats four times.

Table 1 shows the student's results.

Sample number	Number of mayfly nymphs in $100 \, cm^3$ of pond water
1	2
2	3
3	7
4	0
5	8

Table 1

1–4 Calculate the mean number of mayfly nymphs per m^3 of pond water.

$1 \, m^3 = 10^6 \, cm^3$ *[2 marks]*

1–5 The pond is a circular shape with a diameter of $3 \, m$ and a depth of $0.5 \, m$.

Calculate the number of mayfly nymphs in the pond.

Use the formula:

area of a circle $= \pi r^2$

where $\pi = 3.14$ and $r = $ radius

Give your answer in standard form.   *[4 marks]*

1–6 Some fertiliser from a field is washed into the pond.

Explain why the fertiliser could cause an increase in the number of mayfly nymphs
in the pond.   *[2 marks]*

Total: 12

2   The whole human genome has now been studied and will have great importance in the future.

2–1   Define the term **genome**.   *[1 mark]*

2–2   Give **one** reason why it is important to understand the genome.   *[1 mark]*

        The genetic material of a cell is composed of a chemical called DNA.

        DNA is contained in structures found in the nucleus of a cell.

2–3   Name the structures that contain DNA.   *[1 mark]*

        Figure 2 shows part of a DNA molecule.

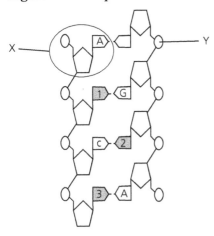

        Figure 2

2–4   Name parts **X** and **Y** on **Figure 2**.   *[2 marks]*

2–5   Give the letters of the complementary bases in the shaded boxes, 1–3 on **Figure 2**.   *[1 mark]*

2–6   Describe the structure of DNA.   *[2 marks]*

        Figure 2 shows that DNA has a specific sequence of bases.

2–7   Explain what could happen if an **A** was changed to a **G**.   *[2 marks]*

                                                                                     Total: 10

3   Tropisms are plant growth responses.

        The plant growth response to gravity is called geotropism (gravitropism).

3–1   Give **one** other example of a plant growth response.   *[1 mark]*

3–2   Name the plant growth hormone that causes geotropism.   *[1 mark]*

3–3 A student investigates the effect of gravity on the growth of a newly germinated seedling.

The student grows a seed in a box for two days then turns the box on its side.

Figure 3 shows the seedling inside the box.

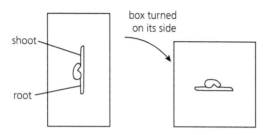

Figure 3

The student leaves the box on its side for three days.

Draw a diagram to show the appearance of the seedling after three days. Label the shoot and the root.   *[2 marks]*

3–4 Explain how the plant growth hormone named in question 3–2 causes the root's response to gravity.   *[4 marks]*

3–5 Explain how gravitropism helps a plant to grow.   *[3 marks]*

Total: 11

4   Accommodation is the ability of the human eye to change its focus between near and distant objects.

Figure 4 shows a human eye focused on a near object.

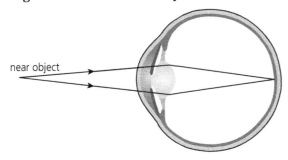

near object

Figure 4

4–1   Explain how the eye changes to focus on a distant object.   *[6 marks]*

4–2   Some people have myopia (short sightedness).

Explain why someone with myopia cannot focus clearly on distant objects.   *[2 marks]*

4–3   The blink reflex is a protective response.

In the blink reflex, the eyelid closes if an object touches the cornea.

Describe the path taken by the nerve impulse in the blink reflex.   *[3 marks]*

Total: 11

5 The fruit fly has either long or short wings.

Figure 5 shows the inheritance of wing length in fruit flies.

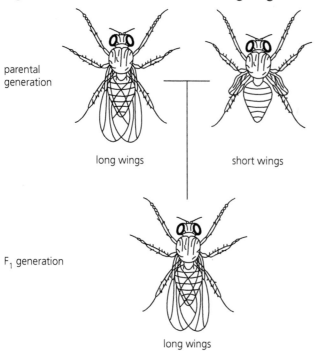

Figure 5

5–1 Identify the dominant allele for wing length. *[1 mark]*

Use the following symbols for your answers to 5–2 and 5–3:

D = dominant allele for wing length

d = recessive allele for wing length

5–2 Give the genotype of a fruit fly in the F_1 generation. *[1 mark]*

5–3 Two flies from the F_1 generation are crossed to give the F_2 generation.

Some of the F_2 generation have short wings.

Draw a Punnett square genetic diagram to show why some of the F_2 generation have short wings.

Give the phenotypes of all of the offspring. *[3 marks]*

5–4 Give the percentage of the offspring that have short wings. *[1 mark]*

5–5 Fruit flies can be cloned and used to study genetics.

Figure 6 shows how a fruit fly can be cloned.

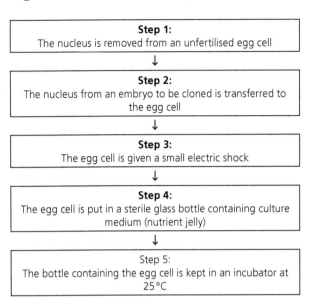

Figure 6

Give a reason for each of the following steps in fruit fly cloning.　　　*[5 marks]*

- The nucleus is removed from the egg cell
- The egg cell is given an electric shock
- The egg cell is put in a sterile bottle
- The bottle contains culture medium (nutrient jelly)
- The egg cell is kept at 25 °C

5–6 The fertilised egg cell in **Step 1** shown in **Figure 6** is taken from a fly with long wings. The embryo in **Step 2** shown in **Figure 6** is taken from a fly with short wings.

Give the reason why the egg cell from **Step 5** in **Figure 6** develops into a fly with short wings.　　　*[1 mark]*

5–7 **Table 2** shows the classification of three species of fly.

Group	Fruit fly	Fungus gnat	Spotted wing drosophila
Kingdom	Animalia	Animalia	Animalia
Phylum	Arthropoda	Arthropoda	Arthropoda
Class	Insecta	Insecta	Insecta
Order	Diptera	Diptera	Diptera
Family	Drosophilidae	Sciaridae	Drosophilidae
Genus	Drosophila	Sciara	Drosophila
Species	melanogaster	hemerobioides	suzukii

Table 2

Explain what the table shows about the evolutionary relationship between the three species of fly.　　　*[2 marks]*

Total: 14

6 The kidneys maintain water balance in the body.

6–1 Name the hormone that acts on the kidney to control water balance in the body. *[1 mark]*

6–2 Name the gland that releases the hormone involved in helping the kidneys to control water balance. *[1 mark]*

6–3 A change in blood concentration can cause water loss from the body to be reduced.

Explain how the kidneys reduce water loss from the body. *[3 marks]*

6–4 People who suffer from kidney failure may be treated by an organ transplant or by using kidney dialysis.

Kidney dialysis needs to be carried out three times per week and each dialysis session lasts about three hours. Before dialysis can start, the person needs to have an operation to create a special blood vessel in their arm called an AV fistula. Dialysis is not painful, but some people can feel sick or dizzy and may have muscle cramps. People on dialysis will have severe restrictions on the amount of fluid they can drink and the amount of minerals in their diet.

Kidney transplants require a donor kidney and the average waiting time for a donor kidney is two and a half to three years. A kidney transplant is a major surgical procedure with a number of risks. Most people can leave hospital about one week after the transplant. People who have had a transplant will need to take immunosuppressant medication for the rest of their lives.

Evaluate the treatments for kidney failure.

Use your own knowledge and information from the question. *[5 marks]*

Total: 10

7   Bacteria are important in the decay of biological material.

Bacteria reproduce at a fast rate.

A scientist investigates the optimum conditions for bacterial growth.

Figure 7 shows the oxygen uptake for a population of bacteria over 24 hours.

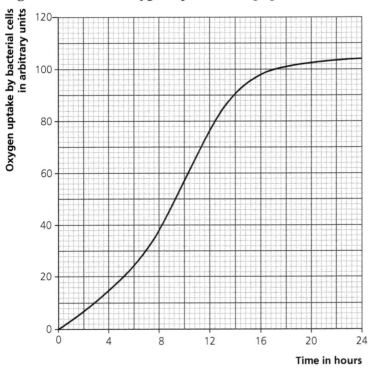

Figure 7

7–1 Give the reason why the oxygen uptake increases between 0 and 4 hours.   *[1 mark]*

7–2 The rate of oxygen uptake is faster at 10 hours than at 16 hours.

Calculate how many times faster the rate of oxygen uptake is at 10 hours than at 16 hours.

Use **Figure 7** to find the rates at 10 and 16 hours.   *[4 marks]*

7–3 The rate of reproduction in the bacterial population decreases after 16 hours.

Suggest **three** reasons why the reproduction rate decreases.   *[3 marks]*

Total: 8

8   Maize is a cereal crop used to produce food for humans and other animals.

8–1 Describe an investigation to estimate the size of a population of maize in a field.   *[4 marks]*

8–2 Maize plants can be genetically modified to be resistant to a herbicide called glufosinate sodium.

Explain the advantage of making maize resistant to glufosinate sodium.   *[3 marks]*

Total: 7

9   **Figure 8** shows changes in a woman's hormone levels and body temperature during the menstrual cycle.

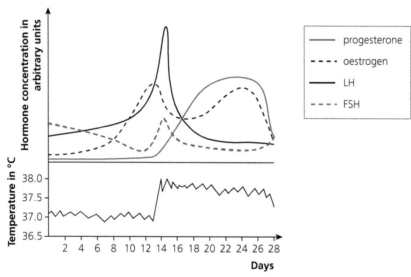

Figure 8

9–1 Explain the interactions of FSH, LH, oestrogen and progesterone in the control of the menstrual cycle.   *[5 marks]*

9–2 Explain how a woman could use measurements of her body temperature to find the time of the month that she is most likely to get pregnant.   *[3 marks]*

9–3 Some couples have unexplained infertility. This means that there is no biological reason why they cannot get pregnant.

Clomifene citrate is a fertility drug used to stimulate ovulation.

Scientists investigated the effectiveness of clomifene citrate compared to other treatments for infertility.

The scientists randomly allocated 576 women to one of three groups for six months.

- **Group 1** (192 women) were given the fertility drug clomifene citrate and were given advice about measuring temperature to find the best time of the month to get pregnant.

- **Group 2** (191 women) were given artificial insemination where sperm were transferred into the uterus through a small tube after ovulation had occurred.

- **Group 3** (193 women) were not given any treatment or advice on measuring temperature to find the best time to get pregnant.

To get valid results, the scientists matched the three groups for as many factors as possible.

Suggest **three** factors that the scientists should have matched.   *[3 marks]*

9–4 The scientists recorded the live birth rate, pregnancy rate, miscarriage rate and multiple birth rate for each group. They also monitored side effects of the treatments.

The study was published in the peer-reviewed British Medical Journal.

Give **one** reason why the results were peer-reviewed before they were published. *[1 mark]*

9–5 The main outcome that the scientists looked at was the live birth rate. They also monitored adverse effects including reports of headaches and nausea.

Table 3 shows the scientists' results.

	Treatment		
	Group 1 – clomifene citrate and advice	**Group 2 – artificial insemination**	**Group 3 – no treatment or advice**
Number of live births	29	43	33
p value compared to Group 3	0.49	0.18	–
Number of reports of headaches	33	4	6
Number of reports of nausea	22	3	4

Table 3

The p value is used to determine whether there is a significant difference between the results.

- A p value ≤ 0.05 shows the difference between the two results is significant.
- A p value > 0.05 shows that the difference between the two results is **not** significant.

A news report on the study had the headline:

FERTILITY TREATMENTS 'NO BENEFIT'

Evaluate the headline.

Use information from **Table 3**. *[5 marks]*

Total: 17

Total for Paper 2: 100

Answers

Cell structure (p. 1)

Quick questions

1 A cell that contains a nucleus.
2 Has genetic material not enclosed in a nucleus
3 Plant cells are eukaryotic.
4 Bacterial cells are prokaryotic.
5 Cellulose
6 Form different types of cells / become specialised.
7 The resolution/resolving power is the shortest distance between two points or lines that can be distinguished.
8 Ribosomes are too small OR light microscope magnification is not high enough.
9 Microscopy techniques have developed over time, including the use of electron microscopes.
10. magnification = size of image ÷ size of real object
11 real object size = image size ÷ magnification; image size = real object size × magnification
12 1×10^{-3} mm; 1×10^{-6} μm; 1×10^{-9} m
13 area = πr^2
14 4.56×10^5; 3.2×10^{-4}

Exam-style questions

15–1 Cell A [1]
15–2 *Any* [2] *marks from:*
Cell A has no nucleus OR Cell B has a nucleus [1]; Cell A has no mitochondria OR Cell B has mitochondria [1]; Cell A is smaller (than cell B) OR Cell B is larger (than cell A) [1]; Cell A has flagella [1]; Cell B has a vacuole [1]; Cell B has chloroplasts [1]
15–3 Cell A = 0.002 mm [1] = 2000 nm [1]
15–4 2 (μm) × 40 000 = X (μm) × 400; 80 000 ÷ 400 = X [1]; so, X = 200 (μm) [1]
15–5 (smallest) ribosome, mitochondrion, nucleus [1]
15–6 Mitochondria are too big to fit inside prokaryotic cells OR prokaryotic cells are too small to contain mitochondria [1].
16–1 *Award* [1] *mark for each correct column [maximum 3].*
Prokaryotic cells only: Plasmid [1]; **Eukaryotic cells only:** Nucleus [1]; **Prokaryotic and eukaryotic:** Cell membrane, Cell wall, Cytoplasm [1].
16–2 *Any* [2] *marks from:*
Prokaryotic genetic material not in nucleus OR eukaryotic genetic material in nucleus [1]; Prokaryotic cells have plasmids OR eukaryotic cells do not have plasmids [1]; Prokaryotic cells have DNA in a loop OR DNA is not in a loop in eukaryotic cells [1]; Prokaryotic cells have a single loop of DNA OR eukaryotic cells have several (linear) chromosomes [1].
16–3 *Award* [1] *mark for each correctly matched structure [maximum 5].*
1-D [1]; 2-A [1]; 3-B [1]; 4-E [1]; 5-C [1]
16–4 Plant cells have a (cellulose) cell wall OR animal cells do not have a cell wall [1]; Plant cells have a permanent vacuole OR animal cells do not have a permanent vacuole [1].
16–5 Electron microscopes have a much higher magnification [1] and a greater resolving power [1].
16–6 It acquires different sub-cellular structures [1] so it is adapted to carry out a certain function [1].
16–7 Most types of animal cells differentiate at an early stage, [1] whereas many plant cells are able to differentiate throughout life [1].
16–8 *Answer should include the structure and function of at least three cells. To get full marks, both plant and animal cells should be included.*
Sperm cell (animal) function is to fertilise an egg cell (ovum) so it has a tail to swim to ovum OR large numbers of mitochondria to provide energy for swimming [1]. Nerve cell (animal) function is to conduct electrical impulses so it has a long axon to conduct impulses over long distances OR myelin sheath to insulate and speed up nerve impulse [1]. Muscle cell (animal) function is to contract and cause movement so it has many mitochondria to provide energy for contraction [1]. Root hair cell (plant) function is to absorb water from the soil by osmosis and minerals by active transport so it has a large surface area for rapid diffusion OR many mitochondria to provide energy [1]. Xylem cell (plant) function is to transport water and mineral ions from roots to other parts of the plant so it has dead cells so water does not move into cells OR continuous tubes so uninterrupted flow of water OR lignin in walls for strength [1].
Phloem cell (plant) function is to transport sugar made during photosynthesis from leaf to other parts of the plant so it has few cell organelles so solution can flow more easily OR elongated cells can form tubes [1].
17–1 A = eyepiece lens [1]; B = objective lenses [1]; C = stage [1]; D = fine focus [1]; E = coarse focus [1]
17–2 Gently rub a cotton bud inside your cheek to collect some cells then rub the cotton bud on the surface of a glass slide [1]. Add one drop of (methylene blue) stain [1] and lower a glass coverslip onto the slide [1].
17–3 Put the slide on the stage [1] and select the low power (smallest) objective lens [1]. Look at the specimen through the eyepiece lens [1] and adjust the coarse focus until the image is clear [1]. Change the objective lens to the high power (larger) lens [1], then use the fine focus to produce a clear image [1].
17–4 Drawing has sharp clear lines with no sketching or shading, and looks like the cell in the photo [1]. The cell membrane, cytoplasm and nucleus are correctly labelled [1].
17–5 Cell membrane – Controls what enters and leaves the cell [1]; Cytoplasm – Site of most chemical reactions in a cell [1]; Nucleus – Contains the genetic material of the cell [1].
17–6 *Any* [2] *marks from:*
has no cell wall OR only has a cell membrane [1]; has no permanent vacuole [1]; has no chloroplasts [1].
17–7 Turn the focusing knob until the image is clear [1].
17–8 Change the objective lens OR rotate the nosepiece [1] to a higher power lens [1].
18–1 Binary fission [1]
18–2 Enough nutrients [1] and a suitable temperature [1].
18–3 Liquid culture medium called nutrient broth OR Solid culture medium called agar jelly (in a Petri dish where bacteria grows as colonies) [1].
18–4 The Petri dish and agar jelly need to be sterilised before using them so there are no unwanted bacteria [1]. The student needs to sterilise an inoculating loop by passing it through a flame to kill any bacteria

on the loop [1]. The sterile loop can be used to collect bacteria from the liquid broth and then spread the liquid on the solid agar jelly [1]. The lid of the Petri dish should be lifted up as little as possible so that bacteria from the air do not enter and contaminate the culture [1]. The student should secure the lid of the Petri dish with four pieces of tape so that the lid cannot fall off and let bacteria enter [1]. Finally, the Petri dish should be incubated upside down (so that condensation does not drip onto the colonies) and at 25 °C so the bacteria can grow (but not potential pathogens) [1].

18–5 mean division time = 40 minutes and 24 hours = 1440 minutes; number of divisions = 1440 ÷ 40 = 36 [1]; Population doubles every 40 minutes, so population size after 36 divisions = 2^{36} [1] = 6.87×10^{10} [1]; *Accept* 6.9×10^{10} *or* 6.872×10^{10}

18–6 diameter = 12 mm, so radius = 6 mm using πr², area of zone = 3.14×6^2 = 113 mm² [2]; *Award* [1] *mark if diameter used instead of radius 452.16 mm²* [1].

18–7 A [1] because it has the largest zone of inhibition OR because it kills the most bacteria [1].

Cell division (p. 5)

Quick questions

1 Nucleus
2 DNA
3 Large numbers of genes
4 Two (found in pairs)

Exam-style questions

5–1 A – Nucleus [1]; B – chromosome [1]; C – gene [1]
5–2 Growth/repair [1]
5–3 It can differentiate into any type of plant cell, throughout the life of the plant [1].
5–4 Clones of plants can be produced quickly [1] and economically [1].
5–5 Crop plants with useful characteristics such as disease resistance can be cloned [1] to produce large numbers of identical plants for farmers [1].
6–1 DNA replication (to form two copies of each chromosome) [1].
6–2 The cell grows OR the number of ribosomes/mitochondria/sub-cellular structures increases [1].
6–3 8 [1]
6–4 D [1]
6–5 One set of chromosomes is pulled to each end of the cell [1] and the nucleus divides.

6–6 The cytoplasm and cell membrane have divided [1] to form two identical (daughter) cells [1].
7–1 An undifferentiated cell that can give rise to many more cells of the same type, [1] and can differentiate to produce other cell types [1].
7–2 Stem cells from human embryos – Can be cloned and made to differentiate into most different types of human cells [1]; Stem cells from adult bone marrow – Can form many types of cells including blood cells [1].
7–3 The cells are not rejected by the patient's body [1].
7–4 Diabetes [1] and paralysis [1]
7–5 It has potential risks; for example, the transfer of viral infection [1]. Some people have ethical or religious objections; for example, using human embryos is destroying a potential life [1].
7–6 *Any* [1] *mark from each of the four sections below, plus* [1] *additional mark for any other point below. Answer must include advantages and disadvantages of each type of stem cell.*
Advantages of using stem cells from the patient are:
• There are no ethical issues as the patient can give consent [1].
• It is a safe procedure for the patient [1].
• It is a reliable technique that has been tried and tested over many years [1].
Disadvantages of using stem cells from the patient are:
• There is a risk of infection associated with having surgery [1].
• The procedure can be painful, so this might put patients off [1].
Advantages of using stem cells from human embryos are:
• If the embryos have already been created, it is better to use them than waste them [1].
• Cells from embryos can differentiate into any human cell [1].
• Using embryos would not cause pain to the patient [1].
Disadvantages of using stem cells from human embryos are:
• It is a relatively new technique, so there may be side effects or long term problems [1].
• There could be a risk of transferring viral infection, or of the stem cells dividing uncontrollably [1].

• The embryo could be a potential human life that is destroyed to provide stem cells [1].
• The embryo is not able to give consent to the procedure [1].
Answer needs to have a conclusion for the final [1] *mark:*
In conclusion, it is better to use stem cells from human embryos as they are more readily available, don't cause pain to anyone, and can differentiate into any type of human cell. OR In conclusion, it is better to use stem cells from the patient since there are no ethical issues, the procedure is more reliable and there is no risk of viral infection [1].

Transport in cells (p. 8)

Quick questions

1 Diffusion is the spreading out of the particles of any substance in solution, or particles of a gas, resulting in a net movement from an area of higher concentration to an area of lower concentration.
2 The cell membrane
3 Into cells = *any two from:* glucose / amino acids / oxygen; Out of cells = *any two from:* carbon dioxide / lactic acid / urea
4 Two of roots, stem or leaves.
5 Osmosis is the diffusion of water from a dilute solution to a concentrated solution through a partially permeable membrane.
6 Only lets certain substances/molecules through.
7 percentage change in mass = (change in mass ÷ starting mass) × 100
8 Active transport moves substances from a more dilute solution to a more concentrated solution (against a concentration gradient) using energy from respiration.
9 They absorb them (into root hair cells) from the soil.
10 Active transport
11 Sugar / glucose

Exam-style questions

12–1 *Calculation of surface area and volume for two cubes with side lengths that double:*
6 cm² and 1 cm³ and 24 cm² and 8 cm³ OR 24 cm² and 8 cm³ and 96 cm² and 64 cm³ [1]
Calculation of surface area to volume ratio for two cubes with side lengths that double:
6 and 3 OR 3 and 1.5 [1]
As the length of the side doubles surface area to volume ratio is halved [1].

12–2 A single-celled organism has a relatively large surface area to volume ratio whereas a large multicellular organism has a relatively small surface area to volume ratio [1].

12–3 Their large surface area to volume ratio [1] allows oxygen to diffuse efficiently into the cells through the cell membrane [1].

12–4 Insects are small and so have a relatively large surface area to volume ratio [1] and can obtain their oxygen by diffusion [1].

12–5 Mammals have an exchange surface / breathing system where oxygen diffuses into the blood [1] and a transport system / circulatory system to take the oxygen to the body cells [1].

13–1 Temperature [1] and the surface area of the membrane [1].

13–2 A steeper concentration gradient (bigger difference in concentration) [1] will cause a faster rate of diffusion [1] because the particles will move more readily from higher to lower concentration [1].

13–3 Network of capillaries give rich blood supply [1] Ventilation (breathing) provides supply of fresh air [1]

13–4 Walls of villi are only one cell thick [1] so short diffusion path [1]. OR Network of capillaries/rich blood supply [1] so steep concentration gradient [1]. OR Millions of villi [1] so large surface area [1].

13–5 Any [2] marks from:
Many gill filaments so large surface area [1]; Good blood supply so maintains a steep concentration gradient [1]; Thin wall / one cell thick so short diffusion distance [1].

14–1 So that only water could move into and out of the bag OR So that sugar could not move into or out of the bag [1].

14–2 0.9 ÷ 10.7 × 100 [1] = (–)8.4% [1]

14–3 Water moves out of the bag by osmosis [1] from a more dilute to a more concentrated solution [1].

14–4 Tube A – Will gain mass [1] because water will move into the bag by osmosis from a more dilute to a more concentrated solution [1]; Tube B – Mass will stay the same [1] because same concentration inside and outside the bag so no net movement of water by osmosis [1].

15–1 Concentration of salt solution [1].

15–2 Change in mass of carrot cylinders [1].

15–3 Any [2] marks from:
Same volume of each solution [1]; Same diameter/length of carrot cylinder [1]; Cylinders in solution for same length of time [1].

15–4 So water on the carrot does not increase the mass [1].

15–5 Plot a graph with Salt concentration on the x-axis and Percentage change in mass on the y-axis [1]; Draw a line of best fit through the points [1]; Find the concentration where the line crosses the x-axis OR find the concentration where there is no change in mass [1].

16–1 (Both are) passive processes (do not require energy) [1]; (Both) involve movement of substances down a concentration gradient [1].

16–2 (Active transport) requires energy from respiration [1]; (Active transport) moves substances against a concentration gradient [1]; Accept converse for osmosis.

16–3 Cell has a large surface area [1] for diffusion/osmosis [1].

16–4 Active transport [1]

16–5 (Aerobic) respiration [1] happens in the mitochondria. This releases energy [1] to move the mineral ions against their concentration gradient [1].

16–6 Less oxygen [1] so less aerobic respiration [1] and less energy for active transport OR less energy to transport mineral ions against their concentration gradient [1].

Cell biology topic review (p. 12)

1–1 Water entered the potato (cells) by osmosis [1] from a more dilute solution outside the cells to a more concentrated solution inside the cells [1].

1–2 –13.5 % [1]

1–3 Potatoes had different starting masses OR calculating percentage change allows comparisons of the results [1].

1–4 Concentration on x-axis, Percentage change on y-axis, suitable scales chosen [1]; All points correctly plotted [1]; Smooth curve drawn through points [1]; Both axes fully labelled including units [1].

1–5 Correct reading from graph where line of best fit crosses the x-axis [1].

1–6 It is the same as the concentration of the solution inside the potato cells [1].

2–1 A – mitochondrion [1]; B – (permanent) vacuole [1]; C – cell membrane [1]

2–2 Protein synthesis [1]

2–3 It has a nucleus [1].

2–4 ×556 [2] Allow [1] mark for incorrect conversion of units, for example, ×0.1 or ×1

2–5 Select a larger/bigger/higher power objective lens [1] then adjust the (fine) focus [1].

3–1 Mitochondria are the site of (aerobic) respiration [1]. More mitochondria will provide more energy [1] for the active transport of glucose [1].

3–2 Microvilli increase the surface area of the membrane [1] so the rate of diffusion increases [1].

3–3 Stem cells from human embryos can differentiate into almost any type of human cell [1]; Cells differentiate to carry out a certain function OR cells differentiate to become specialised cells [1]; The specialised cells divide by mitosis to produce more of the same type of cells [1].

4–1 They have a small surface area to volume ratio [1] so they need specialised exchange surfaces [1].

4–2 Any [2] marks from:
Large surface area [1]; Thin / one cell thick / short diffusion path [1]; Good blood supply / efficient blood supply / rich in blood capillaries [1].

4–3 Stem cells [1]

4–4 72 picograms [1]; 36 picograms [1]

4–5 Diabetes OR paralysis OR blindness [1]

4–6 Cruel to animals OR may not work in humans OR animals cannot give consent [1]

Principles of organisation (p. 15)

Quick questions

1 Cells

2 A group of cells with a similar structure and function.

3 An aggregations of tissues performing a specific function.

Exam-style questions

4–1 (Smallest) Cell → Tissue → Organ → Organ system → Organism (Largest) [2]; Award [1] mark for three correct

4–2 Any [2] marks from:
muscle, epithelial, glandular, epidermal, palisade mesophyll, spongy mesophyll, xylem, phloem, meristem (Accept any other tissue)

4–3 Any [2] marks from:
brain, heart, lungs, liver, kidneys, leaf, root (Accept any other organ)

4–4 Any [2] marks from:
nervous system, circulatory system, digestive system, plant transport system (Accept any other system)

Animal tissues, organs and organ systems (p. 15)

Quick questions

1   To digest and absorb food.
2   Digestion – the breakdown of large, insoluble food molecules into smaller, soluble molecules; Digestive enzymes – carbohydrase, protease, lipase.
3   Enzymes – biological catalysts that speed up chemical reactions in all living things. All enzymes are proteins. Enzymes have an active site with a specific shape that the substrate fits into.
4   An organ that pumps blood around the body in a double circulatory system.
5   It has two circuits (pulmonary and systemic) so the blood goes through the heart twice.
6   Found in the wall of the heart / in the heart muscle tissue where they supply glucose and oxygen to the cells of the heart.
7   People with problems with their pacemaker can have an artificial pacemaker fitted (an electrical device used to correct irregularities in the heart rate).
8   Blood consists of: Plasma – (straw-coloured) liquid that transports dissolved glucose, amino acids, urea and carbon dioxide; Red blood cells – transport oxygen (as oxyhaemoglobin); White blood cells – defence against pathogens by engulfing pathogens or producing antibodies; Platelets – cell fragments that help blood to clot.
9   A stent is a wire mesh that can be inserted into a coronary artery to keep it open so more blood can flow through.
10 An unhealthy lifestyle can cause a build-up of fatty material in the coronary arteries and reduces the supply of oxygenated blood to the heart, eventually causing a heart attack. The risk of coronary heart disease can be reduced by eating a balanced diet, not smoking, exercise and taking statins (drugs that can be used to reduce blood cholesterol levels).
11 Health is the state of physical and mental well-being. Health can be improved by: balanced diet, regular exercise, reduced stress.
12 Risk factors are linked to an increased rate of a disease; for example, poor diet, alcohol or smoking can increase the risk of type 2 diabetes and cancer.

Substances in a person's body or environment; for example, carcinogenic chemicals or ionising radiation can increase the risk of cancer.
13 Obesity means very overweight (with a BMI greater that 25).
14 A carcinogen is a chemical or agent that causes cancer by damaging DNA and causing mutations, for example, tar in cigarettes, asbestos, UV, X-rays.
15 Cancer is the result of changes in cells that lead to uncontrolled growth and division. Malignant tumours are formed and they can spread from one part of the body to another.

Exam-style questions

16–1 A – oesophagus [1]; B – stomach [1]; C – pancreas [1]; D – small intestine / ileum [1]; E – large intestine / colon [1]; F – liver [1]
16–2 1C [1]; 2D [1]; 3A [1]; 4B [1]
16–3 Put four test tubes into a test tube rack and add a sample of liquid food to each test tube.
Carbohydrate (reducing sugar): Add Benedict's solution to the food then stand the tube in a water bath of hot water (80 °C) [1]. If the colour changes from blue to brick-red then sugar is present [1]. OR Carbohydrate (starch): Add iodine solution [1] to a tube and if the colour changes from orange-brown to blue-black, then there is starch in the food [1].
Lipid: Add ethanol to a tube and shake thoroughly then add water [1]. If the solution turns milky white, then the food contains lipid [1].
Protein: Add Biuret solution [1] to the final tube and a colour change from blue to purple shows that there is protein present [1].
16–4 Digestive enzymes convert food into small soluble molecules that can be absorbed into the bloodstream [1].
16–5 **A:** Starch [1]; **B:** Pancreas [1]; **C:** Fatty acids and glycerol [1]; **D:** Protein [1]; **E:** Amino acids [1]
16–6 Bile is alkaline to neutralise hydrochloric acid from the stomach [1] and provide the optimum pH for lipase [1] Bile emulsifies fats (breaks large drops of fat into small droplets) [1]. This increases the surface area for lipase to act on and speeds up the rate of lipid breakdown [1].
17–1 The enzyme has a specific shape and has an active site specific to one substrate [1]; The substrate fits into the active site like a key

in a lock [1]; The enzyme catalyses the reaction and a bond is broken [1]; The products are released from the active site and the enzyme is unchanged [1].
17–2 As temperature increases from 0 °C to 41 °C, the rate of reaction increases [1]. This is because the enzyme and substrate molecules have more kinetic energy, so more successful collisions take place [1]. The enzyme works best at 41 °C, the enzyme's optimum temperature [1]. Above 41 °C, the rate of reaction starts to decrease [1] because the enzymes start to denature [1]. This is when the active site changes shape and an enzyme is permanently damaged [1].
17–3 Time taken for all starch to break down OR Time taken for no further colour change [1].
17–4 *Any* [1] *mark from:* volume of amylase [1]; volume of buffer [1]; volume of starch [1]; time in water bath [1]; number of drops of iodine per well [1]; number of drops of mixture removed every 30s [1].
17–5 Using a water bath [1].
17–6 Temperature affects the rate of an enzyme-controlled reaction. An increase in temperature would increase the rate [1]. If too cold, rate of reaction will be low OR If too hot, enzymes can denature [1].
17–7 Gives a more accurate measurement of the amount of starch OR Gives a value closer to the true value OR It is a quantitative measure whereas colour change is subjective [1].
17–8 Remove drops of the mixture more frequently; for example, every 10 or 20 seconds [1].
17–9 The enzyme denatures at the low/acidic pH [1]. The active site changes shape so the substrate will not fit [1].
18–1 A – trachea [1]; B – lung [1]; C – bronchus [1]; D – alveoli [1]; E – capillary network [1]
18–2 Chemicals (called carcinogens) in cigarette smoke [1] damage DNA [1] causing cells to divide uncontrollably [1].
18–3 *Any* [2] *marks from:* asthma [1]; bronchitis [1]; emphysema [1].
18–4 *Any* [2] *marks from:* Chemicals in cigarette smoke increase heart rate / blood pressure [1]; Chemicals in cigarette smoke

damage the lining of the arteries [1]; Fatty material builds up in the (coronary) arteries [1]; (Coronary) arteries are narrowed [1]; Supply of oxygenated blood to the heart muscle is slowed/stopped [1].

18–5 *Any* [2] *marks from:*
Increased risk of miscarriage [1]; Increased risk of premature birth [1]; Increased risk of low birth weight [1].

19–1 A – Vena cava [1]; B – Pulmonary artery [1]; C – Aorta [1]; D – Left atrium [1]; E – Left ventricle [1]

19–2 The 'pacemaker' is a group of cells located in the right atrium [1] that controls the natural resting heart rate [1].

19–3 Left ventricle – pumps blood around the body [1]; Right ventricle – pumps blood to the lungs where gas exchange takes place [1].

19–4 *Any* [3] *marks from:*
high fat diet [1]; smoking [1]; high blood cholesterol [1]; lack of exercise [1]

19–5 In coronary heart disease layers of fatty material build up inside the coronary arteries [1] narrowing them and reducing the flow of blood [1] resulting in a lack of oxygen for the heart muscle [1].

20–1 The fault may prevent the valve from opening fully OR the heart valve might develop a leak OR the valve might not close properly [1].

20–2 Less oxygenated blood would be pumped around the body [1] so the person may feel tired/dizzy/breathless [1].

20–3 Faulty heart valves can be replaced using biological/mechanical valves [1].

20–4 Artificial hearts are used to keep patients alive while they are waiting for a heart transplant, or to let the heart rest to help it recover [1].

20–5 *Any* [2] *marks from:*
The transplant may be rejected [1]; Patient needs to take immunosuppressant drugs to prevent rejection [1]; Waiting lists for donor can be very long [1]; There is a risk of infection – viral infection from the donor heart, or infection from the operation [1].

21–1 A – Artery [1]; B – Vein [1]; C – Capillary [1]

21–2 Arteries have a thick layer of muscle tissue and elastic tissue to maintain and withstand high pressure OR narrow lumen to maintain high pressure [1]. They transport blood away from the heart at high pressure [1]. Veins have a wide lumen so less resistance to blood

flow OR valves to prevent backflow of blood [1]. They transport blood to the heart at low pressure [1]. Capillaries have thin walls (only one cell thick) so substances can diffuse into and out of the blood [1]. They exchange materials between blood and cells/tissues [1].

22–1 A – red blood cells/erythrocytes [1]; B – platelets/thrombocytes [1]; C – white blood cells/lymphocytes/phagocytes [1]; D – plasma [1]

22–2 1C [1]; 2A [1]; 3D [1]; 4B [1]

22–3 *Any* [2] *marks from:*
Glucose [1]; Amino acids [1]; Lactic acid [1]; Mineral ions [1]; Vitamins [1]; Hormones [1]

22–4 Contain the red pigment haemoglobin [1] to bind to oxygen [1]. OR No nucleus [1] so more room for haemoglobin [1]. OR Biconcave disc shape [1] so large surface area to volume ratio [1].

23–1 Communicable diseases are infectious, non-communicable diseases are not infectious. OR Communicable diseases are caused by pathogens, non-communicable diseases are not caused by pathogens. OR Communicable diseases can be transmitted from one person to another, non-communicable diseases cannot be transmitted [1].

23–2 *Any* [2] *marks from:*
Diet [1]; Stress [1]; Life situations (such as bereavements, money issues, and so on) [1]

23–3 1B [1]; 2A [1]; 3D [1]; 4C [1]

23–4 Causation – There is a causal mechanism that has been proven for some risk factors [1], for example, chemicals in cigarette smoke have been shown to **cause** cancer in laboratory animals. Correlation – There is a relationship between two variables, but this does not mean that one causes the other [1].

23–5 Cardiovascular disease [1]; Cancer [1]

23–6 Negative correlation / as percentage abdominal fat increases insulin sensitivity decreases [1].

23–7 *Any* [2] *marks from:*
Graph shows a correlation between percentage abdominal fat and insulin sensitivity, not that the percentage abdominal fat causes diabetes [1]; The sample size is small OR only 22 people in the sample [1]; Do not know what level of insulin sensitivity is linked with type 2 diabetes OR what level of insulin sensitivity is normal [1].

24–1 Liver can become fatty and enlarged [1]. Long term use can cause liver cirrhosis OR the liver becomes scarred and cannot function properly [1].

24–2 Short term effects – slower reaction times OR loss of balance OR slurred speech [1]; Long term effects – memory problems OR anxiety OR depression OR reduced brain function [1].

24–3 High levels of alcohol can cause fetal alcohol syndrome OR increased risk of miscarriage OR premature birth OR low birth weight OR reduced brain function [1].

24–4 *Any* [2] *marks from these human costs:*
increased crime [1]; violence [1]; antisocial behaviour [1]; death or injury from drink driving [1]
Plus any [2] *marks from these financial costs:*
expense to NHS of treatment for injuries or alcoholism [1]; loss of earnings due to absence from work / inability to work [1]

25–1 Ionising radiation, such as X–rays and some UV light [1]; Chemicals, such as tar in cigarette smoke, asbestos [1]

25–2 Tumours form through uncontrolled growth and cell division [1].

25–3 Malignant tumours invade neighbouring tissues, benign tumours do not OR Malignant tumours spread to different parts of the body in the blood, benign tumours do not OR Malignant tumours form secondary tumours, benign tumours do not [1].

25–4 (Malignant) tumours can spread through the blood / via the circulatory system [1].

25–5 *Any* [3] *marks from:*
Smoking [1]; Too much alcohol [1]; Ionising radiation [1]; Obesity [1]

Plant tissues, organs and systems (p. 22)

Quick questions

1   Meristem tissue is found at the growing tips of roots and shoots. It contains stem cells that divide (by mitosis) to produce new cells for growth (of the root and shoot).

2   Transpiration. The rate is increased by more wind, high temperature, less humidity and high light intensity.

3   The movement of water from roots to leaves in the xylem: water enters root hair cells by osmosis then travels from cell to cell through the root. Water enters the xylem and travels up the stem to the leaves.

4   Transpiration: Provides water for cells to keep them turgid; Provides water to the leaf cells for photosynthesis; Transports mineral ions from roots to leaves.

5   Allow oxygen and carbon dioxide to diffuse into and out of the leaf. Mainly found on lower surface of a leaf.

6   Guard cells

7   The movement of food molecules (dissolved sugars) made in the leaves to the rest of the plant through phloem tissue.

Exam-style questions

8–1   Organ [1]

8–2   A – cuticle [1]; B – upper epidermis [1]; C – palisade mesophyll layer [1]; D – spongy mesophyll layer [1]; E – lower epidermis [1]; F – guard cell [1]; G – stoma [1]

8–3   1B [1]; 2D [1]; 3E [1]; 4C [1]; 5A [1]

8–4   Epidermal tissue is transparent so that light can pass through easily [1]. Palisade mesophyll tissue is made of cells that have a regular shape and are tightly packed to absorb more light for photosynthesis OR Palisade mesophyll tissue is made of cells that contain large numbers of chloroplasts to absorb more light for photosynthesis [1]. Spongy mesophyll tissue is made of irregularly–shaped cells with lots of air spaces between them so that gases can diffuse rapidly [1]. Xylem tissue is composed of hollow tubes strengthened by lignin so they can transport water in continuous columns [1]. Phloem tissue is composed of tubes of elongated cells and their end walls have pores so sap can easily move from one cell to the next [1]. Meristem tissue contains stem cells which can differentiate into other types of cells allowing growth [1].

9–1   Roots, stem and leaves [1]

9–2   needed for photosynthesis OR transports substances in solution OR turgidity of cells [1]

9–3   Root hair cells take up mineral ions by active transport [1] against the concentration gradient [1] (from low concentration in the soil to higher concentration in the cells). Aerobic respiration in the mitochondria [1] provides the energy [1] required for active transport.

9–4   Water evaporates from the leaves [1]. Water moves out through the stomata [1] and this creates a pull [1] moving water up through the plant through the xylem from roots to leaves [1].

9–5   High temperature – water molecules have more kinetic energy and evaporate more quickly [1]; High light intensity – stomata are open in bright light so more water can diffuse out [1]; High air movement – water vapour is carried away from the leaf so there is a low concentration of water vapour outside the stomata / there is a steep concentration gradient for rapid diffusion [1]; Low humidity – less water vapour in the air means there is a steep concentration gradient for rapid diffusion out of the leaf [1].

10–1   X – stoma/pore [1]; Y – guard cell [1]

10–2   To prevent water loss and wilting [1].

10–3   There is a high concentration of potassium ions inside guard cells so water enters the guard cells [1] by osmosis [1]. The guard cells swell OR the guard cells become turgid [1] and stomata open because the outer wall stretches more than the inner wall [1].

10–4   Area of field of view = πr^2 = 3.14 × 0.2^2 = 0.1256 [1]; seven stomata in 0.1256 mm^2 [1]; 7 × (1 ÷ 0.1256) stomata in 1 mm^2 = 55.73 = 56 to the nearest whole number [1].

10–5   Count the number of stomata in more fields of view (a minimum of five) [1] and then calculate the mean number of stomata [1].

10–6   Put the leaf on a piece of cm squared paper and draw round the leaf [1]. Count the number of squares inside the leaf outline [1].

10–7   Lower surface has lower temperature [1] so less water lost by evaporation/transpiration [1].

Organisation topic review (p. 25)

1–1   Blood = Tissue [1]; Blood, blood vessels and heart = Organ system [1]; Heart = Organ [1]

1–2   *Clear line diagram with no sketching or shading* [1]; *Two of cell membrane, cytoplasm or nucleus labelled.* [1]; *Magnification given* [1].

1–3   An artery has thicker layers of muscle tissue and elastic tissue than the vein [1]. The lumen of an artery is narrower than the lumen of a vein [1]. An artery does not have valves, but a vein does [1].

1–4   40 cm^3 in 1 second; 40 × 60 × 60 cm^3 in 1 hour = 144 000 cm^3 [1]; 1 cm^3 = 0.001 dm^3, so 144 000 cm^3 = 144 dm^3 [1]

1–5   Contraction of left ventricle [1]

1–6   Decrease = 40 to 0.03 cm^3 s^{-1} = 39.97 cm^3 s^{-1}; percentage decrease = 39.97 ÷ 40 × 100 = 99.925 % [1] = 99.9% (to 3 significant figures) [1]

1–7   More time for exchange of materials by diffusion [1] between blood and body cells [1].

2–1   The digestive system has several organs that work together to digest and absorb [1] food. Digestive enzymes convert food into small soluble [1] molecules that can enter the blood stream. Enzymes catalyse specific reactions in living organisms due to the shape of their active site [1]. Enzymes are denatured [1] at high temperatures.

2–2   *Award* [1] *mark for each correct column.*
Amylase: Salivary gland; Pancreas [1]; **Lipase:** Pancreas [1]; **Protease:** Stomach; Pancreas [1].

2–3   Add Benedict's reagent [1] and heat [1]. A brick-red colour (precipitate) shows that sugars are present [1].

2–4   Fatty acids and glycerol [1]

2–5   Bile is not released into the small intestine [1] so lipids cannot be emulsified OR broken down into small droplets [1] and there is a smaller surface area for lipase to act on [1]. OR Bile is not released into the small intestine [1] so hydrochloric acid from the stomach is not neutralised [1] and the pH is too low for lipase to work efficiently [1].

2–6   *Any* [3] *marks from:*
Percentage of obese females and males has increased since 1994 [1]; Higher percentage of obese females than obese males [1]; Difference between percentage of obese females and males has reduced [1]; Number of obese males has almost doubled [1]; Number of obese females has increased by 8.5% [1]

2–7   People are less active OR do less physical work OR drive more [1]; People eat more food/fat/fast food/processed food [1].

3–1   So that water cannot evaporate from the flask OR so that any water loss is through the leaves [1].

3–2   *Time on x-axis and water loss on y-axis with suitable linear scales* [1]; *Both axes labelled including units* [1]; *All points plotted accurately* [1]; *Straight lines used to join points* [1].

3–3   Rate of water loss between 0 and 4 hours = 0.4 [1]; Rate of water loss between 4 and 12 hours = 1.25 [1]; Unit is g hour^{-1} OR grams per hour [1].

3–4 The rate of transpiration was slower between 0 and 4 hours [1] because there was lower temperature OR higher humidity OR less air movement OR lower light intensity [1]. OR The rate of transpiration was faster between 4 and 12 hours [1] because there was higher temperature OR lower humidity OR more air movement OR higher light intensity [1].

4–1 Alcohol intake below about 40 g per day reduces the risk of CHD in women OR Alcohol intake above about 40 g per day increases the risk of CHD in women OR Alcohol intake below about 95 g per day reduces the risk of CHD in men OR Alcohol intake above about 95 g per day increases the risk of CHD in men [1].

4–2 Any [3] marks from:
Graph only shows effect of alcohol on CHD, but brain/liver/unborn babies can also be affected [1]; Factors other than alcohol intake could cause the decrease in CHD [1]; No information about the age of people in the study, or the number of people involved [1]; Data may not have been peer reviewed [1]; No information about the amount of alcohol in one alcoholic drink [1]; Correlation does not mean there is a causal link [1].

4–3 Rate of blood flow through the coronary artery of a person with CHD would be slower/reduced [1] because layers of fatty material build up inside the coronary arteries [1] narrowing them [1] and reducing the flow of blood.

4–4 Taking statins can cause serious side effects, but these are rare [1] and statins do reduce the risk of heart attack or stroke [1]. However some people will still have a heart attack or stroke even though they are taking statins [1].
High blood cholesterol is only one risk factor for heart disease; if people lead an unhealthy lifestyle, they could still have a heart attack or stroke [1]. People may rely on statins and not make healthy life style choices once they start taking them [1].
In conclusion, people should try to reduce their risk of serious heart conditions by making lifestyle changes rather than relying on statins [1].

Communicable disease (p. 30)

Quick questions

1 Pathogens are microorganisms that cause infectious disease.

2 Pathogens may be viruses, bacteria, protists or fungi.

3 Virus

4 HIV attacks the body's immune cells OR white blood cells

5 Antiretroviral drugs

6 Many species of plants

7 Bacteria

8 Bacteria

9 Purple or black spots develop on leaves. The leaves often turn yellow and drop early. Plant growth is affected because photosynthesis is reduced.

10 Fungus

11 Protists / Plasmodium (eukaryotic microorganisms)

12 The immune system.

13 Antibiotics are medicines that help to cure bacterial disease by killing (infective) bacteria inside the body.

14 A medicine that treats the symptoms of disease, but does not kill pathogens.

15 Alexander Fleming

16 Tablet or other medicine with no drug in it OR Something that looks like the real drug, but contains no drug.

17 A trial where neither the doctor nor the patient know if they are getting the real drug or a placebo.

Exam-style questions

18–1 Any [3] marks from:
Coughs and sneezes from infected people create droplets in the air that can be breathed in by other people OR Spores from fungi are very small and can be blown through the air [1]; Some animals are vectors for disease OR mosquitoes can spread disease when they bite people OR aphids can spread disease when they feed from plants [1]; Contaminated food that has been prepared in unhygienic conditions OR not cooked properly can contain high numbers of pathogens [1]; An infected organism can spread disease to an uninfected organism by direct contact / when they touch / by sexual contact / when leaves of plants touch [1]; Water can be contaminated with pathogens and can infect people when they drink it [1].

18–2 Skin: The skin forms a continuous barrier OR the outer layer of skin is made of dead cells that are difficult to penetrate OR Glands secrete an oily substance called sebum that kills/inhibits the growth of microorganisms [1].
Hairs in nose: Hairs around the nostrils act as a filter for inhaled air removing dust and pathogens [1].

Ciliated epithelium in trachea and bronchi: Pathogens stick to mucus and then the cilia move the mucus away from the lungs towards the throat to be swallowed [1].
Hydrochloric acid in the stomach: The acid in the stomach kills any pathogens ingested in food and drink, or in swallowed mucus [1].

18–3 White blood cells destroy pathogens by phagocytosis [1]. This is when a white blood cell (phagocyte) engulfs a pathogen and then destroys it [1] by digesting it. Other types of white blood cells (lymphocytes) produce chemicals called antibodies [1]. Antibodies have specific shapes and only bind to and destroy specific pathogens [1]. Lymphocytes also produce chemicals called antitoxins [1] to counteract toxins (poisons) [1] produced by some pathogens.

19–1 Bacteria can reproduce rapidly inside our body [1] and may produce poisons (toxins) that damage tissues [1] and make us feel ill.

19–2 By bacteria ingested in food OR on food prepared in unhygienic conditions [1].

19–3 Any [2] marks from:
fever [1]; abdominal cramps [1]; vomiting and diarrhoea [1]

19–4 The salmonella bacteria secrete toxins (poisons) [1] that damage tissues and make us feel ill [1].

19–5 By vaccinating poultry OR Wash hands thoroughly before and after handling food OR Clean kitchen surfaces before and after preparing food OR Avoid eating undercooked eggs, beef, pork and especially poultry [1].

20–1 Gonorrhoea is spread by sexual contact OR It is a sexually transmitted disease (STD) [1].

20–2 Thick yellow or green discharge from the vagina or penis and pain on urinating [1].

20–3 By treatment with antibiotics OR using a barrier method of contraception such as a condom [1].

20–4 Different antibiotics work in different ways [1], so specific antibiotics will kill specific bacteria [1].

20–5 Many strains of gonorrhoea bacteria have developed resistance to antibiotics [1].

21–1 Viruses may reproduce rapidly inside the body [1]. Viruses live and reproduce inside cells, causing cell damage [1].

21–2 HIV is spread by sexual contact OR exchange of body fluids e.g. through the blood when drug users share needle [1].

21–3 HIV initially causes a flu-like illness [1].

21–4 The body's immune system becomes so badly damaged it can no longer deal with cancers or with infections [1].

21–5 Antibiotics cannot kill viral pathogens OR Antibiotics are only effective against bacteria because viruses live inside cells [1].

21–6 Viruses live and reproduce <u>inside cells</u> [1], so drugs that kill viruses may also <u>damage body tissues</u> [1].

22–1 Fever [1] and red skin rash [1].

22–2 By inhalation of droplets [1] from coughs and sneezes [1].

22–3 A vaccine is a small quantity of <u>dead</u> or <u>inactive</u> pathogen [1].

22–4 Measles is a serious illness that can be fatal if complications arise [1].

22–5 A small amount of <u>dead</u> or <u>weakened</u> measles virus [1] is injected into the body. The <u>white blood cells</u> (lymphocytes) [1] produce <u>antibodies</u> [1] specific to the virus. If the measles virus enters the body, the white blood cells <u>rapidly</u> [1] produce <u>large</u> numbers of antibodies [1] that destroy the pathogen and prevent disease [1].

23–1 A distinctive pattern of discolouration on the leaves that looks like mosaic [1].

23–2 Purple or black spots develop on leaves. The leaves often turn yellow and drop early [1].

23–3 <u>Photosynthesis</u> is reduced [1], so <u>less glucose</u> is produced [1] and plant growth is stunted/reduced OR plants cannot synthesise the proteins needed for growth [1].

23–4 Using fungicides (chemicals that kill the pathogen) [1]. Removing and destroying the affected leaves [1].

24–1 Through the bite of a mosquito [1].

24–2 Malaria causes recurrent episodes of fever [1] and can be fatal [1]. *Allow other suitable symptoms such as headache, nausea, abdominal pain.*

24–3 Mosquitoes lay their eggs on still water, so draining ditches and pools [1] prevents mosquitoes from breeding [1].
Mosquitoes spread malaria when they bite, so using mosquito nets over beds at night OR wearing insect repellent [1] prevents mosquitoes from biting [1].

25–1 1B [1]; 2C [1]; 3A [1]

25–2 Aspirin – Painkiller [1]; Penicillin – Antibiotic [1]

25–3 Traditionally drugs were extracted from plants and microorganisms [1], whereas most new drugs are synthesised by chemists in the pharmaceutical industry [1].

25–4 Toxicity – To check the drug is not <u>poisonous</u> or harmful [1]; Efficacy – To check that the drug <u>works</u> to treat the disease [1]; Dose – To check the <u>optimum</u> amount to give [1].

26–1 (Pre-clinical testing uses) cells OR tissues OR live animals [1].

26–2 To check that the drug is <u>effective</u> (that it works) OR to check that the drug is <u>safe</u> [1].

26–3 To check for <u>side effects</u> or unexpected reactions to the drug OR to check that the drug is <u>safe</u> [1].

26–4 Patients may be taking other drugs OR side effects may be symptoms of the disease rather than the new drug [1].

26–5 To find the <u>optimum</u> dosage [1].

26–6 For <u>comparison</u> with the group being given the drug, to see whether the drug works (rather than any improvement being due to the placebo effect) [1].

26–7 Peer review involves other scientists looking at the data to check that claims are <u>valid</u> and to detect <u>false claims</u> [1].

Monoclonal antibodies (p. 33)

Quick questions

1 Identical antibodies produced from a single clone of cells.

2 Lymphocytes

3 Lymphocytes produce antibodies specific to protein antigens.

4 Tumour cell / Myeloma cell

5 Hybridoma cell

Exam-style questions

6–1 Monoclonal antibodies are <u>specific</u> to one binding <u>site</u> on one protein <u>antigen</u> and so are able to <u>target</u> a specific <u>chemical</u> or specific cells in the body.

6–2 antigen [1]; lymphocytes [1]; hybridoma [1]; cloned [1]; antibody [1]

6–3 *Any* [2] *marks from:*
In pregnancy tests [1]; To measure the levels of hormones/chemicals in blood [1]; To detect pathogens [1]; To identify specific molecules in a cell or tissue [1]. To target specific molecules for treatment [1].

6–4 Urine moves through the reaction site [1] and the mobile <u>antibodies</u> bind to the HCG <u>hormone</u> [1] in the pregnant woman's urine. When the urine reaches the results site, the HCG hormone binds to the immobilised antibodies [1]. The antibodies also have blue dye attached so a line appears in the results site to show that the woman is pregnant [1]. The urine travels up to the control site where mobile antibodies from the reaction site bind to the immobilised antibodies [1] and the blue dye on the mobile antibodies gives a line to show that the test is complete [1].

7–1 Monoclonal antibodies can be bound to radioactive substances/toxic drugs/chemicals which stop cells growing and dividing [1]; The monoclonal antibodies will target cancer cells without harming other cells in the body [1].

7–2 The scientist injects the protein/receptor/HER2 into a mouse [1] to stimulate the mouse <u>lymphocytes</u> to produce <u>antibodies</u> [1]. These lymphocytes would be collected and combined with <u>tumour</u> cells [1] to form a <u>hybridoma</u> [1]. The hybridoma cell would then be <u>cloned</u> [1] and these clones divide rapidly and produce antibodies [1].

7–3 Herceptin/MABs have a specific shape [1] so they bind to the receptor/antigen [1] on the surface of cancer cells [1].

7–4 Herceptin is a targeted drug therapy so it only affects cancer cells and not healthy cells [1]; Herceptin causes many side effects, but can be used as a treatment for advanced breast and stomach cancer which could be fatal [1]; MABs, such as Herceptin are made using mice, so some people may object to the use of animals that might suffer pain [1]; Herceptin cannot be used for all types of breast and stomach cancer as some do not have large amounts of HER2 [1].

7–5 Monoclonal antibodies create more side effects than expected [1].

Plant diseases (p. 36)

Quick questions

1 Virus

2 Fungus

3 Aphids / spider mites / white flies

4 Nitrate

5 Magnesium

Exam-style questions

6–1 *Any* [3] *marks from:*
stunted growth [1]; spots on leaves [1]; areas of decay (rot) [1]; growths on stems or leaves [1]; malformed stems or leaves [1]; discolouration e.g. yellow leaves (chlorosis) [1]; presence of pests [1].

6–2 Plant diseases can be identified by referring to a gardening manual or website [1]; taking infected plants to a laboratory to identify the pathogen [1]; using testing kits that contain monoclonal antibodies [1].

6–3 So they can provide the optimum conditions for plants to grow [1].

6–4 Magnesium ions deficiency – chlorosis (yellow leaves) [1]; Nitrate ion deficiency – stunted growth [1].

6–5 The plants would produce less chlorophyll [1] and so can absorb less light [1]. This would reduce the rate of photosynthesis [1] and less glucose would be produced [1]. Less glucose means less protein synthesis and so less growth OR Less glucose means less respiration and so less energy for growth OR Less glucose means less cellulose for new cell walls and so less growth [1] .

7–1 Cellulose cell walls are tough and act as a barrier that microorganisms find difficult to penetrate [1]; Leaves have a waxy cuticle that is tough and difficult for microorganisms to penetrate [1]; A layer of dead cells around stems or the bark on trees falls off taking pathogens with them [1].

7–2 Some plants produce antibacterial chemicals that kill bacteria or inhibit their growth [1]; Some plants produce poisons to deter herbivores [1].

7–3 So that small animals like insects will fall off [1].

7–4 The yellow spots make it look like the leaves already have eggs on them [1], so the butterfly does not lay eggs on the leaves and they are not eaten/damaged by caterpillars [1].

7–5 Prickles can sting/hurt [1] so this prevents animals/herbivores from eating the leaves/fruit/plant [1].

Infection and response topic review (p. 37)

1–1 The measles virus is inhaled in droplets from sneezes and coughs [1]. The virus then sticks to mucus in the trachea and bronchi [1]. Cilia move the mucus away from the lungs OR mucus is moved to the back of the throat then into the stomach [1].

1–2 Reduces the spread of infection OR leads to herd immunity [1]

1–3 Any [2] marks from:
More antibodies are produced after the second vaccination than after the first [1]; Antibodies are produced faster after the second vaccination than after the first [1]; The number of antibodies stays high for longer after the second vaccination [1].

1–4 The side effects of the vaccine are quite mild and complications are rare, whereas the complications of measles infection are more serious [1]. Although the vaccine can cause seizures, the risk of seizures is five times higher if you have the infection [1]. There is a risk of death from getting measles, but no risk of death associated with the vaccine [1]. In conclusion, people should have their children vaccinated against measles as it is a potentially fatal disease that cannot be treated, so it is worth the small risk of side effects that the vaccine can cause [1].

2–1 The yellow patches / lack of chlorophyll mean that less light is absorbed so there will be less photosynthesis [1] and less glucose is made [1]; Less glucose means less respiration [1] so less energy is released for growth [1]. OR Less glucose means less amino acids [1] are produced so less protein can be synthesised for growth [1].

2–2 The scientist injects the antigen into a mouse [1]; The mouse produces antibodies specific to the antigen [1]; The lymphocytes are combined with tumour/myeloma cells to make hybridoma cells [1]; The scientist clones the hybridoma cells to produce many identical cells that make the same antibody [1].

2–3 The oil may be poisonous/harmful to the insect OR The oil may cause the insect to fall off the leaf [1].

2–4 Use pesticide/insecticide to kill the thrips OR Use ladybirds to eat the thrips OR Destroy infected plants OR Remove infected leaves [1].

3–1 Poisons/toxins secreted by the bacteria [1].

3–2 Antibiotics cannot kill viruses [1] because viruses live inside body cells [1].

3–3 The cells of the immune system are damaged [1] so the white blood cells/lymphocytes cannot produce as many antibodies against the Salmonella bacteria [1].

3–4 The new drug must first be tested in pre-clinical trials using cells, tissues or animals [1]. The drug is then tested in clinical trials and the first phase of these uses small numbers of healthy volunteers [1]. The volunteers are given very low doses of the drug to check that it is safe for use in humans [1]. In second phase clinical trials, the drug is tested on small numbers of patients with the disease to check

its efficacy (that it works) [1]. The third phase of clinical trials uses large numbers of patients in double blind trials [1], where the patients are randomly allocated to receive either the real drug or a placebo and neither the doctor nor the patient knows which they are getting [1].

Photosynthesis (p. 39)

Quick questions

1 Photosynthesis is an endothermic reaction in which energy is transferred from the environment to the chloroplasts by light.

2 An endothermic reaction takes in (or absorbs) energy from the surroundings.

3 Chlorophyll

4 carbon dioxide + water
$\xrightarrow{\text{light}}$ glucose + oxygen

5 Carbon dioxide: CO_2; Glucose: $C_6H_{12}O_6$; Oxygen: O_2; Water: H_2O

6 A limiting factor is a factor that directly affects the rate of a reaction, for example, light intensity and carbon dioxide concentration are both limiting factors for photosynthesis. Increasing light intensity will increase the rate of photosynthesis until carbon dioxide becomes the limiting factor and then the rate will level off.

7 Temperature; Light intensity; Carbon dioxide concentration; Amount of chlorophyll

Exam-style questions

8–1 Endothermic [1]

8–2 $6CO_2 + 6H_2O$ [1] $\xrightarrow{\text{light}}$ $C_6H_{12}O_6$ [1] $+ 6O_2$

8–3 Respiration to provide energy [1]; Converted to insoluble starch for storage [1]; Used to produce fat or oil for storage [1]; Used to produce cellulose to strengthen the cell wall [1].

8–4 Set up the equipment as shown in the diagram with the boiling tube a set distance away from the lamp, for example, 20 cm. Put the tank of water between the lamp and the boiling tube to control the temperature [1]. Leave the pond weed for two minutes to adjust to each light intensity before taking any measurements [1]. Count the number of bubbles released from the pondweed in one minute [1]. Repeat this measurements two more times at the same distance so a mean can be calculated [1]. Repeat the measurements at another four distances from the lamp [1], for example, 40, 60, 80 and 100 cm. Use the same piece of pondweed each time [1]. Plot the mean number of bubbles (*y*-axis) against the distance from the boiling tube (*x*-axis) [1].

9–1 Repeat at each distance **OR** use more distances [1].

9–2 The light intensity is greater/higher at 10 cm [1] so more energy is available for photosynthesis [1].

9–3 $V = \frac{4}{3}\pi r^3$; volume of one bubble $= \frac{4}{3} \times 3.14 \times 0.5^3$ [1] $= 0.523$ [1]; four bubbles are released at 40 cm, so volume of gas released at 40 cm $= 4 \times 0.523$ [1] $= 2.09$ [1] mm^3

9–4 At 10 cm, light intensity $= \frac{1}{100} = 0.01$; at 20 cm, light intensity $= \frac{1}{400} = 0.0025$; at 40 cm, light intensity $= \frac{1}{1600} = 0.000625$ [2]; As distance doubles, light intensity is quartered [1].

9–5 One bubble would be released [1] because the distance has doubled so the number of bubbles is quartered [1].

9–6 Another factor has become the limiting factor [1]; for example, temperature/carbon dioxide concentration [1].

9–7 Carbon dioxide concentration: *Description or graph showing*: As carbon dioxide concentration increases, the rate of photosynthesis increases; then (the curve) levels off/plateaus [1]. *Explanation*: Carbon dioxide is one of the reactants/raw materials used in photosynthesis so the rate of photosynthesis increases when more is available [1]. The rate levels off when another factor, such as light intensity, becomes the limiting factor [1].
Temperature:
Description or graph showing: As temperature increases, the rate of photosynthesis increases up to its maximum, then the rate decreases at high temperatures [1]. *Explanation*: The rise in temperature increases the rate of reaction as the molecules have more kinetic energy [1]. The rate of photosynthesis decreases because enzymes denature at high temperatures [1].

10–1 Increasing the carbon dioxide concentration does not increase the rate of photosynthesis [1]. (The curve has levelled off.)

10–2 *Any* [2] *marks from*: temperature [1]; light intensity [1]; amount of chlorophyll [1].

10–3 The tomato grower could grow the tomatoes at 25 °C [1] and at a carbon dioxide concentration of 0.11% [1] because this gives the maximum rate of photosynthesis [1] so the tomatoes would have more glucose for growth/protein synthesis [1].

10–4 It would cost more money than using 0.12% [1] and it would not increase the rate of photosynthesis any further / it would not increase the yield of tomatoes any further [1].

Respiration (p. 41)

Quick questions

1 Cellular respiration is an exothermic reaction which happens continuously in living cells and transfers the energy needed for living processes / releases the energy stored in glucose for living processes.

2 Mitochondrion/Mitochondria

3 Aerobic respiration uses oxygen, and anaerobic happens without oxygen.

4 glucose + oxygen
$\rightarrow$ carbon dioxide + water

5 glucose $\rightarrow$ lactic acid

6 glucose $\rightarrow$ ethanol and carbon dioxide

7 Fermentation

8 Metabolism is the sum of all the reactions in a cell or the body.

Exam-style questions

9–1 $C_6H_{12}O_6$ [1]

9–2 *Any* [2] *marks from*:
chemical reactions to build larger molecules [1]; movement [1]; keeping warm [1]; growth [1]; repair [1]

9–3 The oxidation of glucose is incomplete [1].

9–4 Aerobic respiration uses oxygen, but anaerobic respiration is without oxygen [1]; Aerobic respiration produces carbon dioxide and water, but anaerobic respiration produces lactic acid [1].

9–5 Lactic acid builds up [1] and creates an oxygen debt OR muscles become fatigued OR muscles stop contracting efficiently [1].

9–6 An oxygen debt is the amount of extra oxygen needed after exercise [1] to react with the accumulated lactic acid and remove it from the cells [1].

9–7 The lactic acid is transported from the muscles to the liver [1] where is it converted back to glucose (by an oxidation reaction) [1].

10–1 Breathing rate increases [1] and breath volume increases [1].

10–2 The increase in heart rate supplies more oxygen / more oxygenated blood [1] to the muscles, so more (aerobic) respiration [1] can take place and more energy can be released [1] for muscle contraction.

10–3 Take two groups of people, 10 who exercise once a week and 10 who exercise three or more times a week [1]. Match the groups so that they are the same age and gender [1]. Ask all of the people to sit quietly for five minutes and then find their resting heart rate in beats per minute [1]. Make all of the people do the same exercise for the same length of time, for example, running on a treadmill for three minutes [1]. Measure the heart rate every 30 seconds and record the time it takes to return to the resting rate [1]. Calculate the mean time taken for the heart rate to return to its resting rate for each group, and compare the times for the two groups [1].

11–1 1A [1]; 2C [1]; 3B [1]

11–2 Urea [1]

11–3 The mean metabolic rate of males is higher than females at all ages [1]. The metabolic rates of both females and males decreases with age [1]. The metabolic rates of both females and males decreases rapidly up to about 20 years of age, then decreases more slowly [1].

11–4 Decreases by 36 in 10 years [1]; percentage decrease = 36 ÷ 205 × 100 [1] = 17.5609 = 17.6 to 3 significant figures [1].

11–5 This is when growth is taking place [1] so there is a high rate of respiration [1] to provide the energy [1] needed for growth. OR This is when growth is taking place [1] so there is a high rate of synthesis reactions [1] to provide the proteins [1] for growth.

12–1 Bread making OR beer/wine making [1]

12–2 Carbon dioxide / CO_2 [1]

12–3 To prevent the entry of oxygen [1] so that aerobic respiration does not occur OR so that conditions remain anaerobic [1].

12–4 Count the number of bubbles OR collect the gas and measure the volume [1] in one minute (or other specified time) [1].

12–5 Same volume of yeast [1]; Same volume/concentration of sugar [1].

Bioenergetics topic review (p. 44)

1-1 Carbon dioxide / CO_2 [1]

1-2 Respiration (in the grasshopper) uses <u>oxygen</u> from the air [1] and produces carbon dioxide that is absorbed by the soda lime [1]. This decreases the <u>pressure</u> inside the boiling tube [1] so the bubble moves.

1-3 So the grasshopper is not harmed [1].

1-4 To let fresh air/oxygen in OR to reset the position of the water drop [1].

1-5 Black paper prevents entry of <u>light</u> [1] so that <u>no photosynthesis</u> occurs [1] and <u>only respiration</u> is measured [1].

2-1 Between 12 am and 6 am it is dark / there is no light so there is no photosynthesis [1], <u>only respiration</u> takes place [1].

2-2 From 6 am there was (sun) light so the plants started photosynthesising [1] and the carbon dioxide concentration decreased because the rate of photosynthesis was <u>higher</u> than the rate of respiration [1].

2-3 The rates of photosynthesis and respiration were the same [1], so the amount of carbon dioxide used (in photosynthesis) was the same as the amount of carbon dioxide produced (in respiration) [1].

3-1 *Any* [2] *marks from:*
Sugars/glucose converted to glycogen (for storage in the liver/muscles) [1]; Amino acids used to synthesis proteins (for growth/repair) [1]; Glycerol and three fatty acids used to synthesise lipids (for cell membranes / as energy stores) [1].

3-2 120 (beats per minute) [1]

3-3 4 minutes [1]

3-4 The student's heart rate increased to supply more blood/oxygen/glucose to the muscles [1] for <u>respiration</u> [1] so that more <u>energy</u> can be released [1] for muscle <u>contraction</u> [1].

3-5 *Any* [2] *marks from:*
Run at same speed [1]; Run for same time [1]; Same gender [1]; Same age [1]; Same fitness level [1]

Homeostasis (p. 47)

Quick questions

1 Blood glucose concentration; Body temperature; Water levels

2 Nervous and chemical

3 Receptor cells

Exam-style questions

4-1 Homeostasis is the regulation of the internal conditions of a cell or organism [1] to maintain optimum conditions for function in response to internal and external changes [1].

4-2 All control systems have cells called receptors [1] which detect stimuli/changes in the environment [1]. Coordination centres, for example the brain [1], receive and process information from receptors [1]. Effectors, muscles or glands, [1] then bring about responses which restore optimum levels [1].

4-3 To maintain the <u>optimum</u> temperature [1] for enzyme action [1].

4-4 Glucose is needed for <u>respiration</u> [1] to provide the body with energy/ATP [1].

4-5 Produces urea: liver [1]; Produces urine: kidney [1]; Controls body temperature: skin [1]

The human nervous system (p. 48)

Quick questions

1 Brain and spinal cord

2 Neurones

3 As electrical impulses

4 Brain

5 Light intensity and colour.

6 Accommodation

7 Myopia – short sightedness, and hyperopia – long sightedness

8 The thermoregulatory centre

9 Vasoconstriction – blood vessels constrict/narrow, vasodilation – blood vessels dilate/widen

10 They contract (and relax) involuntarily.

Exam-style questions

11-1 To react to surroundings/external stimuli [1] and to coordinate behaviour [1].

11-2 *Any* [2] *marks from:*
blinking [1]; coughing [1]; pupil constricting [1]; salivating [1]; secreting hormones [1]; regulating heart rate [1]

11-3 To protect the body from danger/damage [1].

11-4 A junction/gap [1] between two neurones [1].

11-5 A <u>chemical/neurotransmitter</u> is released [1] which <u>diffuses</u> across the synapse [1] and attaches to receptors on the next neurone [1].

11-6 The receptor detects the stimulus [1] and generates an electrical impulse [1] that passes along a sensory neurone [1] towards the spinal cord. The impulse crosses a synapse to a relay neurone [1], and across another synapse to a motor neurone [1]. The motor neurone transmits the impulse to a muscle (the effector) which contracts [1] to bring about the response.

12-1 Brain is involved [1]

12-2 2.84×10^{-1} s [1]

12-3 To allow them to identify <u>anomalies</u> OR because there will be <u>variation</u> in the results [1] and to increase the <u>repeatability</u> [1].

12-4 *Any* [2] *marks from:*
Drop the ruler from the same height [1]; Use the same ruler [1]; Use the same hand to catch the ruler [1]; Use same distance between thumb and forefinger [1]; Repeat more times [1].

12-5 The computer will give a more precise reaction time [1]. The students will not be able to tell when the stop sign is about to appear OR the students might be able to tell when the ruler is about to be dropped [1].

12-6 Conclusion is valid because reaction times decrease for students A and C [1].
Plus any [2] *marks from:*
Conclusion is not valid because student B's mean reaction time has increased [1]; students drank coffee, but we do not know how much caffeine was in the coffee [1]; there might have been other substances in the coffee that affected reaction times, not just caffeine [1]; changes in reaction times were very small OR student C's reaction time only decreased by 7 milliseconds [1]; only one type of reaction has been measured so cannot generalise to all reaction times [1].

13-1 Neurones [1]

13-2 X – cerebral cortex [1]; Y – cerebellum [1]; Z – medulla [1]

13-3 X (cerebral cortex) – Controls conscious thought, behaviour, language, intelligence and memory [1]; Y (cerebellum) – Controls balance and movement [1]; Z (medulla) – Controls unconscious activities such as heart rate and breathing rate [1].

13-4 The brain is very complex [1] and very delicate [1].

13-5 Electrically stimulating different parts of the brain [1] and using MRI scanning techniques [1].

13-6 This allows scientists to find out more about the different parts of the brain and their functions [1]. If the patient is brain damaged they may not be able to give consent to the procedure OR the techniques could cause more damage to the patients' brains [1].

14–1 X – cornea [1]; Y – retina [1]; Z – optic nerve [1]

14–2 To protect the eye [1].

14–3 The radial muscles in the iris contract [1] and the circular muscles relax [1]. This dilates the pupil [1] so that more light can enter.

14–4 The ciliary muscles relax [1] which increases their diameter [1] and pulls the suspensory ligaments tight [1]. The lens is then pulled thinner [1] and refracts light rays less [1] to focus the image on the retina [1].

14–5 The (spectacle) lens refracts (bends) the light [1] so that the light rays focus on the retina [1].

14–6 Hard/soft contact lenses [1]; Laser surgery to change the shape of the cornea [1].

15–1 Brain [1]

15–2 Receptors [1] in the skin detect the temperature and send (electrical) impulses [1] along sensory neurones [1] to the brain/thermoregulatory centre.

15–3 Sweat is produced from the sweat glands [1] and blood vessels dilate [1].

15–4 Both (sweating and vasodilation) cause a transfer of energy [1] from the skin to the environment [1]. OR Sweat on the skin evaporates, using heat energy from the skin to do so [1]. Vasodilation results in more blood being carried to the skin surface where its heat can be lost to the air [1].

15–5 Muscles contract (and relax) rapidly [1] which increases the rate of respiration so more heat energy is released [1].

Hormonal coordination in humans (p. 51)

Quick questions

1 Glands

2 Chemical secreted by the endocrine system.

3 In the blood.

4 At their target organ.

5 Pituitary gland

6 Pancreas

7 Insulin

8 Glycogen

9 In liver and muscle cells

10 Through the lungs during exhalation, from the skin in sweat, through the kidneys as urine.

11 Water, ions and urea

12 Kidneys

13 Testosterone produced in the testes

14 Oestrogen produced in the ovaries

15 Follicle stimulating hormone/FSH so no eggs mature

16 One of: condoms, diaphragms

17 Kill or disable sperm

18 FSH and LH

19 Adrenaline/follicle-stimulating hormone (from adrenal glands) and thyroxine/luteinising hormone (from thyroid gland)

Exam-style questions

20–1 Gland [1]

20–2 A – pituitary gland [1]; B – thyroid gland [1]; C – adrenal gland [1]; D – pancreas [1]

20–3 A hormone is a chemical messenger [1] produced by an endocrine gland [1] and secreted directly into the blood(stream) [1]. The blood carries the hormone to a target organ [1] where it produces its effect.

20–4 The pituitary gland secretes several hormones [1] which act on other glands [1] to stimulate other hormones [1] to be released and bring about effects.

20–5 The adrenal glands produce adrenaline [1] in times of fear or stress [1]. It increases the heart rate and boosts the delivery of oxygen and glucose to the brain and muscles [1], preparing the body for 'flight or fight'.
The thyroid gland produces thyroxine [1] which stimulates the basal metabolic rate [1] and plays an important role in growth and development [1].

20–6 The low level of thyroxine would be detected [1] by receptors and the body would respond by stimulating the thyroid gland to secrete more thyroxine to bring the level back to normal [1]. This is an example of negative feedback [1].

21–1 The person ate a meal containing carbohydrate / drank a sugary drink [1].

21–2 Pancreas [1]

21–3 The pancreas released insulin [1] so that glucose moved into (liver) cells [1] for storage (as glycogen).

21–4 If the blood glucose concentration starts to fall, the pancreas releases glucagon [1]. Glycogen is converted back to glucose and released into the bloodstream [1].

21–5 Type 1 diabetes – the pancreas fails to produce sufficient insulin [1]. Type 2 diabetes – the body cells no longer respond to insulin produced by the pancreas [1].

21–6 Type 1 diabetes – is normally treated with insulin injections [1]. Type 2 diabetes – is commonly treated with a carbohydrate controlled diet and an exercise regime [1].

21–7 Blood glucose level would rise more/higher [1] and would take longer to return to normal [1].

21–8 $(3.7 \times 10^6 \div 6.5 \times 10^7) \times 100$ [1] = 5.6923 [1] = 5.7 (%) to 2 significant figures [1].

21–9 $5.7 \div 3.3$ [1] = 1.7 times greater [1].

21–10 Rise in levels of obesity [1].

22–1 Ions [1] and urea [1]

22–2 Water is lost via the lungs during exhalation [1] and from the skin in sweat [1].

22–3 Water would enter cells by osmosis [1] so they do not function efficiently OR so they swell and burst [1].

22–4 ADH increases the permeability [1] of the kidney tubules [1] so more water is reabsorbed back into the blood [1].

22–5 Excess amino acids are deaminated in the liver [1] producing the toxic chemical ammonia [1] which is converted into urea [1] for excretion by the kidneys.

22–6 Glucose is small and is filtered out of the blood [1], but all of the glucose is reabsorbed [1].

22–7 Proteins are too big to be filtered out of the blood [1].

22–8 Patient's blood passes out of the patient's body and through the machine next to dialysis fluid [1]. The blood and dialysis fluid are separated by a selectively permeable membrane [1]. The dialysis fluid has the optimum concentrations of glucose/ions so there is no net diffusion of glucose out of the blood [1]. The dialysis fluid contains no urea, so urea diffuses out of the blood [1].

22–9 Organ transplants are permanent and allow a person to live a normal life [1] but there is a waiting list for organs OR someone has to die for an organ to become available [1] and there is a risk the kidney could be rejected [1]. Dialysis removes urea from the blood and keeps the person alive [1] but is time consuming [1] and there are risks of infection / blood clots [1].

23–1 Ovary [1]

23–2 From 11 arbitrary units on day 1 to 40 arbitrary units on day 12 = 29 arbitrary units in 11 days. 29 arbitrary units in 264 hours = 29 ÷ 264 arbitrary units per hour [1] = 0.110 arbitrary units per hour [1]. *Accept answer with correct follow through if students start on 10 arbitrary units.*

23–3 Ovulation [1]

23–4 FSH causes maturation of an egg in the ovary and stimulates the ovaries to release oestrogen [1]. Oestrogen causes the uterus lining to thicken, inhibits the release of FSH (so that no more eggs begin to mature) [1] and stimulates release of LH from the pituitary gland [1]. LH stimulates the release of the egg and inhibits oestrogen release [1]. Progesterone maintains the thickness of the uterus lining, inhibits the release of FSH [1] and inhibits the release of LH [1].

23–5 The uterus lining breaks down [1] and menstruation/period occurs [1].

24–1 Condom [1]

24–2 Contains hormones that inhibit FSH production [1] so that no eggs mature [1].

24–3 Some people might forget to take a pill OR Hormone level in blood is more constant with implant [1].

24–4 Prevent spread of sexually transmitted disease/STDs/HIV/ gonorrhoea [1].

24–5 As the egg is fertilised but cannot implant, some people may see this as the destruction of a potential human life [1].

24–6 Not easily reversed OR May be permanent OR Risk of infection after operation [1].

25–1 Pituitary gland [1]

25–2 The eggs are fertilised in the laboratory using sperm from the father [1]. The fertilised eggs develop into embryos [1]. One or more embryos are inserted into the mother's uterus (womb) [1] when they are tiny balls of cells.

25–3 Inserting more embryos can lead to multiple births which are a risk to both the babies and the mother [1].

25–4 *Any [2] marks from:*
It is very emotionally and physically stressful [1]; The success rates are not high [1]; There can be increased risks to mother/babies if a multiple birth [1]; It is expensive [1].

Plant hormones (p. 57)

Quick questions

1 Light – phototropism, gravity – geotropism or gravitropism

2 Auxins

3 Gibberellins

4 Ethene

5 Auxins – weed killer / rooting powder / promoting growth in tissue culture.
Ethene – to control ripening of fruit during storage/transport.

Gibberellins – end seed dormancy / promote flowering / increase fruit size.

Exam-style questions

6–1 Shoot grows towards [1] the light [1]. OR Positive phototropism [2].

6–2 Shoot/leaves get more light [1] for photosynthesis [1] so more glucose for energy OR so more glucose for cell wall synthesis OR so more protein for growth [1].

6–3 Plant hormones called auxins [1] accumulate on the shaded side of the shoot [1] and cause cell elongation [1] which makes the shoot curve.

6–4 Answer should include reference to the independent variable (how it could be changed) [1] and the dependent variable (what could be measured) [1]. For example, the student could put the lamp different distances from the box to vary light intensity [1] and could measure the length of each seedling at the start and end of the investigation [1].

6–5 *Any [2] marks from:*
Same number of seeds/seedlings in each plastic dish [1]; Same amount of water on cotton wool [1]; Same type of seed/seedling [1]; Same temperature [1].

7–1 As rooting powders [1]; For promoting growth in tissue culture [1].

7–2 Insects may rely on the weeds for food [1] so killing the weeds may result in loss of some insects which other animals relied on for food, and so biodiversity is decreased [1].

7–3 Bananas are easier to transport when unripe [1] then ethene promotes ripening [1] before the bananas are sold.

7–4 Gibberellins end seed dormancy [1].

7–5 Gibberellins increase fruit/tomato size OR initiate seed germination [1].

Homeostasis and response topic review (p. 58)

1–1 All control systems include cells called receptors which detect stimuli (changes in the environment) [1] and effectors, muscles or glands, which bring about responses [1] to restore optimum levels.

1–2 *Any [3] marks from:*
The endocrine system involves chemicals called hormones and the nervous system involves electrical impulses [1]; Hormones are transported in the blood and electrical impulses travel along neurones [1]; The response by the

endocrine system is slower [1]; The response by the endocrine system lasts for longer [1].

1–3 To focus on a near object the ciliary muscles contract [1], causing the suspensory ligaments to loosen [1] so the lens is thicker and refracts light rays strongly [1].

1–4 The response is a reflex action that does not involve the conscious part of the brain [1].

1–5 Increases heart rate [1] and supplies more oxygen and glucose to the brain and muscles [1].

1–6 The blood vessels [1] supplying the skin will constrict [1] so that less blood flows through the capillaries in the skin [1] and less heat energy will be lost to the surroundings [1]. Shivering will begin when the skeletal muscles begin to contract and relax involuntarily [1]. Respiration is required to provide the energy for muscle contraction and respiration releases heat energy [1] to increase the body temperature.

2–1 volume of water gained by the body = (400 + 1500 = 900) = 2800 cm^3 per day; volume of water lost (in exhaled air, faeces and sweat) = (500 + 150 + 700) = 1350 cm^3 per day; so, volume of water lost in urine = 2800 – 1350 [1] = 1450 cm^3 per day [1].

2–2 A smaller volume of urine will be produced [1] because the hormone ADH [1] will cause more water to be reabsorbed in the kidney tubule [1].

2–3 Sweat evaporates from the skin surface [1] and transfers heat energy away from the skin [1].

2–4 Dilation of the blood vessels supplying the skin capillaries [1] causes more blood to flow near the skin surface [1].

2–5 The low blood glucose level is detected by the pancreas [1] which secretes the hormone glucagon [1] so that stored glycogen is converted into glucose [1].

2–6 Drink should also contain ions [1] to replace those lost in sweat [1].

3–1

	Female	Male
Name	Oestrogen [1]	Testosterone [1]
Site	Ovaries [1]	Testes [1]

3–2 *Any [4] marks from:*
The contraceptive implant is more effective than the progesterone-only pill [1]; The implant lasts for longer than the pill [1]; The implant causes less side effects than the pill [1]; Can be left in place for three years whereas the

pill has to be taken at the same time every day [1]; Can easily be removed if there are side effects / if you want to restore fertility [1]; Will still work even if you have sickness/diarrhoea [1].

3-3 FSH causes more eggs to mature [1] and LH causes ovulation/release of the eggs from the ovaries [1].

3-4 Destroying embryos is destroying a potential human life [1].

3-5 Shoots grow towards light [1] because light is needed for photosynthesis [1]. Roots grow towards the ground [1] to get water and minerals OR to anchor the plant [1].

3-6 Human hormones are transported in blood, but plant hormones are not [1]. Plant hormones diffuse from cell to cell.

Reproduction (p. 61)

Quick questions

1 Meiosis
2 Mitosis
3 Sperm and egg cells
4 Pollen and egg cells
5 Reproduction that only involves one parent and no fusion of gametes.
6 DNA is a polymer made up of two strands forming a double helix.
7 DNA is contained in structures called chromosomes found in the nucleus of a cell.
8 A gene is a small section of DNA on a chromosome.
9 Each gene codes for a particular sequence of amino acids, to make a specific protein.
10 The genome is the complete set of genetic information of an organism.
11 Each nucleotide consists of a common sugar and phosphate group with one of four different bases attached to the sugar.
12 A, C, G and T
13 Three (3)
14 A genetic disorder causing extra fingers or toes.
15 Twenty-three (23)

Exam-style questions

16-1 Copies of the genetic information are made [1] and the cell divides twice to form four gametes [1], each with a single set of chromosomes [1].

16-2 egg [1]; sperm [1]; meiosis [1]; single [1]; fertilisation [1]; mitosis [1].

16-3 The female gamete (egg) and male gamete (sperm) fuse together [1] so the 23 chromosomes from each restore the cell to 46

chromosomes [1].

16-4 The fertilised egg divides by mitosis and the number of cells increases [1]. As the embryo develops, the cells differentiate [1].

16-5 Any [2] marks from:
It is faster than sexual reproduction [1]; It is more energy efficient than sexual reproduction (as no need to find a mate) [1]; It produces many genetically identical offspring when conditions are favourable [1].

16-6 Many fungi reproduce asexually by spores [1] but also reproduce sexually to give variation [1]. OR Many plants produce seeds sexually [1] but also reproduce asexually by runners (in strawberry plants) OR by bulb division (in daffodils) [1].

16-7 In sexual reproduction there is mixing of genetic information [1] that leads to leads to genetic variation [1]. The offspring may be better adapted to survive [1] if the environment changes to become less favourable [1].

17-1 Nucleotides [1]

17-2 X = sugar / deoxyribose [1]; Y = phosphate [1]

17-3 A always links to T (on the opposite strand) and C to G [1]

17-4 The order of bases controls the order in which amino acids are assembled [1]. A sequence of three bases is the code for one particular amino acid [1]. Proteins are synthesised on ribosomes [1] according to a template of the gene [1]. Carrier molecules bring specific amino acids to join to the growing protein chain in the correct order [1]. When the protein chain is complete it folds up to form the specific/unique shape of the protein/enzyme [1].

17-5 The different amino acid could alter the protein so it has a different shape [1]. It changes the shape of the active site [1] so that the substrate can no longer fit/bind [1].

17-6 Non-coding parts can switch genes on and off [1], so variations in these areas of DNA may affect gene expression [1].

17-7 Understanding and treating inherited disorders [1]. Tracing human migration patterns from the past [1].

18-1 A different form of a gene OR A different version of the same gene [1].

18-2 When no dominant allele is present OR When two copies of the recessive allele are present [1].

18-3 No dominant allele present OR Only

recessive allele present [1].

18-4 GG and Gg [1]

18-5

	G	g
g	Gg	gg
g	Gg	gg

Gg = grey fur; gg = white fur [1]. Offspring genotypes = 50% Gg and 50% gg [1]; Probability = 0.5 [1]

18-6 Individual 7 inherited a recessive allele(d) from his father [1] but does not have SCD so must have a dominant allele (D) from his mother [1].

18-7 Parent genotypes = DD and dd [1]; Offspring genotypes = Dd [1]; Offspring phenotypes = unaffected [1]; Probability = 0 [1]

19-1 The man has polydactyly so has the dominant allele [1]. One of the children does not have polydactyly so must have two recessive alleles [1]. The child without polydactyly inherited a recessive allele from each parent [1].

19-2 Boy = XY [1]; Girl = XX [1]

19-3 mother = XX and father = XY [1]; Offspring = XX and XY [1]; Probability = 0.5 [1] (Accept 50%, 1-in-2 or 1:1)

19-4 Parent genotypes = Aa [1]; Parent gametes = A and a [1]; Offspring genotypes = aa, Aa, Aa, aa [1]; Offspring with aa identified as with CF [1].

19-5 Probability = 0.5 [1] (Accept 50%, 1-in-2 or 1:1)

19-6 Any [1] mark from:
Prevent a child from having cystic fibrosis / an inherited disorder [1]; Reduce the long term costs of caring for a child with CF [1]; Prevent future suffering for the child/family [1]; Provide embryos that could be used for stem cells research/treatment [1]; Reduce number of alleles for CF in the gene pool [1].

19-7 Any [1] mark from:
Embryo screening is expensive [1]; Embryos cannot give consent [1]; Involves the destruction of embryos and these are potential human lives [1].

Variation and evolution (p. 64)

Quick questions

1 Differences in the characteristics of individuals in a population.
2 Mutations occur continuously.
3 Very rarely.
4 More than three billion years ago.

5 Selective breeding

6 Tissue culture and cuttings

7 Tissue culture uses small groups of cells from part of a plant to grow identical new plants.

8 Embryo transplants and adult cell cloning.

Exam-style questions

9–1 Variation is due to differences in <u>genes</u> that are inherited (genetic causes) [1] and/or <u>environmental</u> conditions in which an organism develops [1].

9–2 Most mutations have <u>no effect</u> on the phenotype [1], some influence phenotype, but <u>very few</u> determine phenotype [1].

9–3 Very rarely a mutation will lead to a new <u>phenotype</u> [1]. If the new phenotype is suited to a <u>change</u> in the <u>environment</u> [1] it can lead to a relatively rapid change in the species through natural selection [1].

9–4 Evolution is the change in the inherited characteristics of a population [1] over time through a process of <u>natural selection</u> [1].

9–5 The organisms can <u>breed</u> together [1] to produce <u>fertile</u> offspring [1].

9–6 The ancestral population is <u>separated</u>, by a geographical barrier for example, so they cannot interbreed [1]. There is genetic <u>variation</u> in the two populations, or <u>mutations</u> can occur [1]. Each population will experience different <u>environmental</u> conditions [1] and <u>natural selection</u> will occur as some <u>phenotypes</u> are favoured [1]. The <u>alleles</u> for the favourable phenotype are passed on to the offspring [1] and eventually the two populations become so different in phenotype that they can no longer interbreed to produce <u>fertile offspring</u> [1] and two new species have formed.

10–1 *Any* [3] *marks from:*
Disease resistance in food crops [1]; Domestic dogs with a gentle nature [1]; Large or unusual flowers [1]. *Accept other reasonable suggestions.*

10–2 The farmer should choose a female that produces a large volume of milk and a male whose daughters produce a large volume of milk [1] then breed the two animals together [1]. Choose the offspring that have the <u>desired characteristic</u> [1]. Continue over <u>many generations</u> until all the (female) offspring show the desired characteristic [1]. *Accept 'produce a lot of milk' instead of 'desired characteristic'.*

10–3 Some breeds are particularly prone to disease / inherited defects [1].

11–1 *Any* [2] *marks from:*
To produce bigger / better fruits [1]; Increased yield [1]; Herbicide resistance [1]; Drought resistance [1]; Increased nutritional content [1]; Resistance to insect attack [1]; Lack of variation can leave a population vulnerable to changes in environment/disease [1].

11–2 Human insulin OR Blood clotting factors OR Human growth hormone [1].

11–3 To overcome inherited disorders [1]. *Accept answers including a named example of genetic disorder.*

11–4 People may be concerned about the effect GM crops might have on populations of wild flowers and insects [1]. Some people think that the effects of eating GM crops on human health have not yet been fully explored [1].

11–5 <u>Enzymes</u> are used [1] to isolate / 'cut out' the resistance <u>gene</u> from the nematode [1]. This gene is inserted into a <u>vector</u> / bacterial plasmid / virus [1]. The vector is used to insert the gene into the banana plant cells [1]. The genes are transferred to the banana plant cells at an early stage in their development [1] so that they develop with resistance to TR4 [1].

12–1 Small groups of <u>cells</u> are taken from part of a plant [1] and used to grow <u>identical</u> new plants [1].

12–2 To preserve rare plant species OR protect endangered species from extinction [1].

12–3 Cuttings method is simpler OR does not require sterile conditions OR does not require sterile growth medium [1].

12–4 The cells from a developing animal <u>embryo</u> are split apart [1] before they become <u>specialised</u>, then the identical embryos are <u>transplanted</u> into <u>host</u> mothers [1].

12–5 Take an <u>unfertilised egg cell</u> from the sheep and remove its <u>nucleus</u> [1]. Take a body cell from the mouflon and remove its nucleus [1], then put the <u>body cell nucleus</u> into the empty egg cell [1]. Give the egg cell an <u>electric shock</u> [1] so the egg cell starts to divide and form an <u>embryo</u> [1]. Put the embryo in the <u>uterus</u> of an adult female sheep (the surrogate mother) [1].

12–6 *Any* [3] *marks from*
Pros: It prevents the extinction of a species [1]; There may not be enough of them to breed naturally [1]. **Cons:** The success rate for adult cell cloning is very low [1]; Clones are genetically identical so there is less genetic variation [1]; Cloned animals can have developmental problems OR may die earlier than normal [1]; Some people believe it is ethically wrong to clone animals [1]. *Note: answers should include at least one pro and one con to gain all three marks.*
Plus, conclusion needed for final mark: In conclusion, the low success rate and developmental problems are major drawbacks of cloning, but it is worth the risk of these disadvantages to ensure the survival of a species that would otherwise become extinct [1].

The development of understanding of genetics and evolution (p. 67)

Quick questions

1 The theory of evolution by natural selection.

2 On the Origin of Species, published 1859.

3 Alfred Russel Wallace

4 Jean-Baptiste Lamarck

5 Gregor Mendel

6 Fossils are the 'remains' (parts/traces) of organisms from millions of years ago, which are found in rocks.

7 There are no remaining individuals of a species still alive.

8 They reproduce at a fast rate.

Exam-style questions

9–1 Many early forms of life were soft-bodied, which means that they have left few traces behind OR Traces of early life forms have been destroyed by geological activity [1].

9–2 From parts of organisms that have not decayed because one or more of the conditions needed for decay are absent [1], or when parts of the organism are replaced by minerals as they decay [1].

9–3 *Any* [3] *marks from:*
Destruction of habitats [1]; Environmental changes such as ice ages [1]; New predators [1]; New diseases [1]; Competition for food/water/mates/territory [1]; Asteroid collision [1]; Volcanic eruption [1]

9–4 Observations he made on his round the world voyage [1], and years of experimentation and discussion [1].

9–5 Individual organisms within a particular species show a wide range of <u>variation</u> for a characteristic [1]. Individuals with characteristics most suited to the environment are <u>more likely to survive to breed</u> successfully [1]. The characteristics that have enabled these individuals to survive are then <u>passed on</u> to the next generation [1].

9–6 There was insufficient evidence at the time the theory was published to convince many scientists [1]. The mechanism of inheritance and variation was not known until 50 years after the theory was published [1].

9–7 His work on <u>warning colouration</u> in animals OR His theory of <u>speciation</u> [1].

9–8 Changes that occur in an organism during its lifetime [1] can be passed on to their offspring [1].

10–1 Genes [1]

10–2 Chromosomes [1]

10–3 *Any* [2] *marks from:*
The behaviour of chromosomes during cell division was not observed (until the late nineteenth century) [1]; Chromosomes and Mendel's 'units' behaving in the same way was not observed (until the early twentieth century) [1]; The structure of DNA was not determined (until the mid-twentieth century) [1]; The mechanism of gene function was not worked out (until the mid-twentieth century) [1]; Most scientists at the time believed in blended inheritance [1].

11–1 The theory of evolution by natural selection [1].

11–2 <u>Mutations</u> occur in bacterial pathogens and new strains are produced [1]. Some of the new strains are <u>resistant</u> to antibiotics, and so are not killed [1]. The resistant strain <u>survives and reproduces</u>, so the population of the resistant strain rises [1].

11–3 People are not <u>immune</u> to it OR There is <u>no effective treatment</u> [1].

11–4 Doctors should not prescribe antibiotics inappropriately, such as treating non-serious or viral infections [1]. Patients should complete their course of antibiotics so all bacteria are killed and none survive to mutate and form resistant strains [1].

11–5 Bacteria reproduce at a fast rate, so they can evolve rapidly and develop resistance to antibiotics [1]. The development of new antibiotics is very expensive and takes a long time [1].

12–1 **Darwin's theory:** There was natural <u>variation</u> in the size/shape of the finches' beaks [1]. The finches with bigger beaks could eat nuts and get

more food than finches with smaller beaks [1]. The finches with bigger beaks were more likely to survive and <u>breed</u> and pass the gene for bigger beaks to their offspring [1]. **Lamarck's theory:** As the finches ate nuts their beaks got bigger [1]. This bigger beak (acquired characteristic) was passed on to the finches' offspring [1].

12–2 The finch would die and be covered with a layer of sand/silt/mud/sediment [1]. The soft parts of the finch would decay [1] and the hard parts of the finch (bones and beak) would be replaced by minerals [1].

12–3 Fossils would show that the beak shapes have changed over time [1].

12–4 Habitat destruction OR Global warming OR Introducing new predators or pathogens OR Pollution [1].

12–5 Interbreed the two groups of finches [1] and see if they produce fertile offspring [1].

12–6 The populations of finches were <u>isolated/separated</u> from each other [1] by the sea / by a geographical barrier / because they were on different islands [1]. The <u>environmental</u> conditions were different on each island / the <u>food sources</u> were different on each island [1]. Random mutations created variation in the beak shape and size on each island [1] and the phenotypes which were most advantageous in each different habitat thrived [1], passing their alleles to their offspring [1]. Eventually, the populations became so different that they could no longer interbreed [1].

Classification (p. 70)

Quick questions

1 Carl Linnaeus

2 Binomial system

3 Carl Woese

Exam-style questions

4–1 1 = Phylum [1]; 2 = Class [1]; 3 = Order [1]; 4 = Genus [1]

4–2 *Panthera tigris* [1]

4–3 Snow leopard [1]

4–4 Lion and leopard [1]

4–5 The three domains are the <u>archaea</u> which are primitive bacteria [1] usually living in extreme environments, the <u>bacteria</u> which are the true bacteria [1], and the <u>eukaryota</u> which includes protists, fungi, plants and animals [1].

4–6 Improvements in microscopes OR Understanding of biochemical processes OR Genome mapping [1].

Inheritance, variation and evolution topic review (p. 71)

1–1 Child with CF got a CF allele from each parent [1]. Both parents have the CF allele, but do not have the disease [1].

1–2 Parent genotypes = both Hh [1]; Offspring genotypes = HH, Hh, Hh, hh [1]; CF phenotype identified = hh [1]

	H	h
H	HH	Hh
h	Hh	hh

1–3 0.25 / 25% / 1 in 4 / ¼ [1]

1–4 Egg/ovum and sperm [1]

1–5 46 / 23 pairs [1] because gametes join at <u>fertilisation</u> [1] and this restores the chromosome number [1].

1–6 *Any* [3] *marks from:*
Embryo screening would allow the couple to have a healthy child without CF [1]; It also means that the child will not have the CF allele to pass on to their children [1]; Embryo screening is very expensive [1]; It leads to the destruction of embryos which some people think is ethically wrong [1]; Embryos could be damaged by the CF test [1]; There is a risk to the mother during the procedure, for example, infection after the operation [1]. *Note: answers should include at least one pro and one con to gain all three marks.*
Plus, conclusion needed for final mark: Although embryo screening is expensive and there is some risk to the mother, it is worth the risk to have a healthy child who does not carry the CF allele [1].

1–7 Three bases code for <u>one amino acid</u> [1] so the (CFTR) protein will be missing one amino acid [1]. This will change the <u>shape</u> of the (CFTR) protein [1] so the protein can no longer function.

2–1 A new predator appeared OR A new pathogen/disease infected them all OR A new competitor for food/territory OR Asteroid/meteor impact OR Volcanic eruption [1].

2–2 From fossils found in rocks [1].

2–3 There was <u>variation</u> in the Canthumeryx population [1] and those with longer necks were able to reach more leaves/food [1]. The individuals with longer necks were more likely to survive and breed [1] and pass on the genes for a longer neck to their offspring [1].

2–4 Canthumeryx stretched its neck during its lifetime to reach leaves high in the tree [1] and then passed on this characteristic to its offspring [1].

2–5 There was insufficient evidence at the time [1] and the mechanism of inheritance was not known [1].

2–6 About 26–28 million years ago [1].

2–7 They are extinct / they all died out [1] about 21 million years ago [1]. OR They shared a common ancestor with deer [1], about 21 million years ago [1].

2–8 Musk deer and Bovids [1].

2–9 The two populations could have been isolated [1]. There would be genetic variation in each population [1] and the organisms with alleles best suited to the environment [1] are more likely to survive [1] and pass on these alleles to their offspring [1]. After many generations, the two populations can no longer interbreed to produce fertile offspring [1].

3–1 Any [2] marks from: Asexual reproduction is faster than sexual reproduction [1]; Asexual reproduction produces many identical offspring when conditions are favourable [1]; Asexual reproduction is more energy efficient as you don't need to find a mate [1].

3–2 Many early life forms were soft-bodied and did not leave traces [1]. Also, the traces that were left have been destroyed by geological activity [1].

3–3 Archaea [1] and Eukaryota [1]

3–4 Plasmid [1]

3–5 Where chemical reactions take place [1].

3–6 Genetic material/DNA is not enclosed in a nucleus/is free in the cytoplasm [1].

3–7 The scientists could take the plasmid from the bacterial cell [1] and insert the insulin gene into the plasmid [1]. The plasmid is then put back into a bacterial cell and the bacteria are grown [1]. As the bacteria grow they use the insulin gene to make insulin [1] which can be extracted.

3–8 A mutation occurred that made the bacteria resistant to methicillin [1]. The resistant bacteria could not be killed [1] so they survived and reproduced [1] to create a population of resistant bacteria.

3–9 By not prescribing antibiotics for viral infections / for non-serious infections [1] and by telling patients about the importance of completing the course of antibiotics [1].

4–1 Enzymes are used to isolate the required gene from bacteria [1]. The gene is inserted into a vector [1]. The vector is used to insert the gene into the mosquito cells [1]. The gene is transferred to the mosquito cells at an early stage in their development [1].

4–2 Releasing GM mosquitoes into the wild will reduce the population of mosquitoes so there are less mosquitoes biting humans and spreading malaria [1]. Fewer people will be ill with malaria or will die from malaria [1]. This will reduce the cost of controlling the mosquito population and/or of treating malaria [1]. However, developing and producing GM mosquitoes is expensive and the mosquitoes may not survive in the wild [1]. The gene may spread to other insects and cause the death of beneficial insects [1]. There might be effects on the food chain as animals that feed on mosquitoes would have less food [1].

Adaptations, interdependence and competition (p. 74)

Quick questions

1 A group of organisms of the same species that live in the same area at the same time.

2 A group of two or more populations of different species that live in the same area at the same time.

3 Non-living parts of the environment; for example, light intensity, temperature, moisture levels, soil pH and mineral content, wind intensity and direction, carbon dioxide levels for plants, oxygen levels for aquatic animals.

4 Living parts of the environment; for example, availability of food, new predators arriving, new pathogens, one species outcompeting another so the numbers are no longer sufficient to breed.

5 An ecosystem is the interaction of a community of living organisms (biotic) with the non-living (abiotic) parts of their environment; for example, coral reef, desert, marine, rainforest, tundra.

6 Structural, behavioural or functional features that enable organisms to survive in the conditions in which they normally live. Adaptations to extremely cold conditions include: small ears so smaller surface area for heat loss, huddling together to keep warm, or hibernating in winter.

7 Organisms that live in environments that are very extreme, such as at high temperature, pressure, or salt concentration. For example, bacteria living in deep sea vents are extremophiles.

Exam-style questions

8–1 *Award [2] marks for any three correct examples, [1] mark for any two correct examples, and [0] marks for one or zero correct examples, from:* light; space; water; mineral ions.

8–2 food; water; mates; territory [2] *(Allow [1] mark for two correct examples)*

8–3 Each species depends on other species for food, shelter, pollination, seed dispersal, and so on [1]. If one species is removed it can affect the whole community [1].

8–4 All the species and environmental factors are in balance [1] so that population sizes remain fairly constant [1].

8–5 **Community**: all the living organisms present in an ecosystem at a given time, so all of the grass, grasshoppers, lizards, hawks, and so on in the grassland ecosystem [1]; **Ecosystem**: grassland [1]; **Population**: a group of individuals of the same species that occupy the same habitat at the same time; for example, all of the lizards [1].

8–6 (Population would increase because) more grass available for mice to eat [1]; (Populations might decrease because) hawk will eat more mice as no rabbits to eat [1].

9–1 *Any [3] marks from:* light intensity [1]; temperature [1]; moisture levels [1]; soil pH [1]; soil mineral content [1]; wind intensity and direction [1]; carbon dioxide levels for plants [1]; oxygen levels for aquatic animals [1].

9–2 Put two tape measures perpendicular to each other in the shaded area of the field, then use a random number generator [1] to get coordinates. Put a quadrat down at the coordinates [1] then count the number of daisies in the quadrat [1]. Repeat this at least 10 times in the shaded area [1] and find the mean number of daisies per quadrat. Repeat the sampling in the unshaded area of the field [1]. Compare the mean number of daisies in the shaded and unshaded areas [1].

9–3 Light intensity [1] affects the rate of photosynthesis [1]. *(Accept water availability [1] because tree and daisy roots compete for water from the soil [1])*

9–4 20 [1]

9–5 As sulfur dioxide concentration increases, number of different species of lichens decreases [1]; Number of different species of lichen is constant when SO_2 concentration is between 0 and 4 OR no lichen when SO_2 concentration is greater than 180 OR the rate of decrease in the number of species declines sharply above 70 arbitrary units [1].

9–6 Lichen can only grow at low sulfur dioxide concentrations OR Lichen cannot grow at sulfur dioxide concentrations above 180 arbitrary units OR Increased concentration of sulfur dioxide in the air reduces the number of species of lichen growing on trees [1]. *Accept other sensible conclusions.*

10–1 *Any* [3] *marks from:*
Availability of food [1]; New predators arriving [1]; New pathogens/diseases [1]; Competition (one species outcompeting another so the numbers are no longer sufficient to breed) [1].

10–2 percentage decrease = (3 500 000 − 120 000) ÷ 3 500 000 × 100 [1] = 96.6% [1]

10–3 *Any* [3] *marks from:*
The red squirrel is <u>smaller</u> / <u>half the mass</u> of the grey squirrel so it cannot compete as successfully for food [1]; The red squirrel eats a <u>smaller</u> variety/range of food than the grey squirrel so it is more difficult to find food [1]; The red squirrel has a <u>smaller</u> range of habitats available [1]; The same area of land can support a <u>smaller</u> number / <u>five times less</u> red than grey squirrels [1]; The grey squirrel can <u>spread</u> the squirrelpox virus to red squirrels and red squirrels have no <u>immunity</u> [1].

11–1 *Any* [3] *marks from:*
High temperature [1]; High pressure [1]; High salt concentration [1]; High acidity/alkalinity [1].

11–2 Bacteria [1] in deep sea vents [1]. OR Polar bears [1] in the Arctic [1]. OR Emperor penguins [1] in Antarctica [1].

11–3 Their mouth is not <u>damaged</u> by thorns / spines on desert plants [1]; The fat can be used in <u>metabolic</u> reactions, e.g. <u>respiration</u> to produce water [1]; <u>Less water lost</u> in urine so is conserved in the camel's body [1];

11–4 Behavioural [1]

Organisation of an ecosystem (p. 77)

Quick questions

1 Any level in a food chain or web. The first trophic level is almost always a producer (that photosynthesises).

2 Biomass is the mass of living organisms (in a specific area at a specific time). Photosynthetic organisms (green plants and algae) are the producers of biomass.

3 Consumers that kill and eat other animals, for example, fox, lion, shark.

4 Numbers rise and fall in cycles: As the numbers of prey increase, so do the numbers of predators (after a short lag phase). As the numbers of predators increase, the number of prey decrease as they are being eaten. Eventually the reduced number of prey means that the numbers of predators will fall because there is not enough food.

5 Photosynthesis

6 Respiration, combustion

7 Rain, snow, hail or sleet

8 Methane gas

Exam-style questions

9–1 Grass is a <u>producer</u> [1]

9–2 Hare [1]

9–3 Lynx [1]

9–4 Grass produces <u>biomass</u> / <u>synthesises</u> molecules / makes <u>glucose</u> [1] by <u>photosynthesis</u> [1].

9–5 Hare population = 80 000 [1]; Lynx population = 3000 [1]; Hare population is 80 000 ÷ 3000 = 26.7 times greater [1].

9–6 <u>Increased</u> from about 50 000 (in 1910) to about 75 000 (in 1912) [1] then <u>decreased</u> rapidly to less than 1000 (in 1915) [1].

9–7 Population increased because the number of <u>predators</u> was relatively low [1] and then decreased because the number of predators doubled between 1910 and 1912, so more hares were eaten [1].

10–1 The student puts the tape measure straight across the footpath to produce a <u>transect</u> [1] then puts the <u>quadrat</u> at 0 m [1] and <u>counts</u> the number of plantain found in the quadrat [1]. The student then moves the quadrat along the tape measure at 2 m <u>intervals</u> [1] and counts the number of plantain in each quadrat. The student then moves the tape measure along the path to create a new transect parallel to the first [1]

and <u>repeats</u> the sampling at least three times [1].

10–2 Random sampling: to reduce bias [1]; Large number of quadrats: so results are more representative and valid [1].

10–3 Mode = 1 [1]; Median = 1.5 [1]

10–4 Mean = 1.9 [1]

10–5 Area sampled = 400 m² and mean number of plantain per m² = 1.9 × 4 = 7.6 [1]; Total number of plantain = 400 × 7.6 = 3040 [1].

11–1 Temperature [1]; Availability of water [1]; *Allow other reasonable suggestions.*

11–2 Seasonal changes; for example, swallows migrate from Europe to Africa in the winter [1]. Geographic changes; for example, the mosquito that is a vector for malaria is found in tropical and sub-tropical regions of the world, but its distribution could change with global temperature rises [1]. *Allow other reasonable suggestions.*

11–3 A = Precipitation [1]; B = Evaporation [1]

11–4 The loss/evaporation of water from a plant [1] through its leaves/ stomata [1].

11–5 Carbon compounds in the dead plants are broken down by <u>decomposers</u> [1] such as bacteria and fungi. The decomposers <u>respire</u> [1] and release <u>carbon dioxide</u> back into the atmosphere [1] and mineral ions back into the soil. Carbon dioxide diffuses into the leaf of a plant and is used in <u>photosynthesis</u> [1]. Glucose produced in photosynthesis is used to synthesise proteins and cellulose (for cell walls) [1] for <u>growth</u> in the living plant [1].

12–1 *Any* [2] *marks from:*
Higher temperature [1]; More water/moisture [1]; More decomposers/microorganisms [1].

12–2 As a natural <u>fertiliser</u> [1] for <u>growing</u> garden plants/crops [1].

12–3 No oxygen so <u>anaerobic decay</u> takes place [1] producing <u>methane</u> gas that can be used as a fuel [1].

12–4 x-axis: scale in increments of 1 from 0–5 and labelled 'Time in days' [1]; y-axis: scale in increments of 0.5 from 4.5–7.0 and labelled 'pH' [1]; Six points correctly plotted to nearest mm [2]; Smooth curve through points [1].

12–5 Decreases by 1.9 in 120 hours [1]; Rate of decrease = 1.9 ÷ 120 = 0.01583 [1]; In standard form: 1.583×10^{-2} [1]

12–6 Faster at 40 °C because microorganisms <u>reproduce</u> more rapidly at higher temperature [1] so there are <u>more microorganisms</u> to break down the lactose [1]. OR Closer to the optimum temperature for <u>enzymes</u> [1] so lactose is <u>broken down/digested</u> more quickly [1]. Slower at 60 °C because microorganisms/bacteria would be <u>killed</u> by the high temperature [1] so there is <u>no bacteria</u> to break down lactose [1]. OR Enzymes <u>denatured</u> by high temperature [1] so <u>active site</u> is the wrong <u>shape</u> to bind to lactose [1].

Biodiversity and the effect of human interaction on ecosystems (p. 81)

Quick questions

1. The variety of all the different species of organisms on Earth, or within an ecosystem.
2. Sulfur dioxide and nitrogen oxides.
3. Peat is partially decayed vegetation that has been destroyed to produce garden compost, or burnt as a fuel.
4. Cutting down trees (on a large scale) to provide land for cattle and rice fields, and to grow crops for biofuels.
5. The gradual increase in the Earth's average temperature.
6. Carbon dioxide, methane and water vapour.

Exam-style questions

7–1 **Air:** smoke / acidic gases [1]; **Land:** landfill / toxic chemicals [1]; **Water:** sewage / fertiliser / toxic chemicals [1]
7–2 <u>Rapid growth</u> in the population [1] and an <u>increase</u> in the <u>standard of living</u> [1].
7–3 Pollution <u>kills plants and animals</u> [1] which <u>reduces</u> biodiversity [1].
7–4 It reduces the dependence of one species on another [1] for food/shelter/maintenance of the physical environment [1].
7–5 *Any* [2] *marks from:* building [1]; quarrying [1]; mining [1]; dumping waste [1]; destruction of peat bogs [1].
7–6 Improved soil quality increases crop yield [1] and so increases the amount of food produced [1]. Using peat destroys <u>peat bogs</u> OR reduces the area of the <u>habitat</u> [1] and so reduces the <u>biodiversity</u> OR reduces the <u>variety</u> of <u>different</u> plant, animal and microorganism <u>species</u> that live in peat bogs [1]. When

the peat decays, carbon dioxide is released into the atmosphere [1] contributing to global warming [1].
8–1 Percentage of forest lost = 65.41 – 59.05 = 6.36% [1]; Area of forest lost = 540 600 km² [1].
8–2 *Any* [2] *marks from:* To provide land for cattle / rice fields / farming / agriculture [1]; To grow crops for <u>biofuels</u> [1]; To provide wood for building materials [1].
8–3 *Any* [2] *marks from:* Destroys <u>habitats</u> and so reduces <u>biodiversity</u> [1]; Affects the water cycle as there is less <u>transpiration</u> by trees, which means a drier climate [1]; Removes the <u>roots</u> holding soil together and so can cause soil <u>erosion</u> and landslides, and can cause <u>flooding</u> [1]; Reduced water and roots mean soil can turn into desert [1].
8–4 *Any* [2] *marks from:* Melting of polar ice caps/sea ice causing sea levels to rise [1]; Changes in the distribution of species [1]; Extreme weather conditions, such as heavy rainfall leading to flooding or heatwaves leading to droughts [1]; Crops failing OR threat to food security [1].
8–5 Combustion/burning of fossil fuels/wood increases carbon dioxide [1]. Burning/decay of peat increases carbon dioxide [1]. Rice field/cattle increase methane [1].
8–6 Greenhouse gases <u>absorb</u> (heat) <u>energy</u> [1] that is <u>radiated</u> from the Earth's surface [1] and releases this energy back to Earth, keeping the Earth <u>warmer</u> than it would be without the gases [1].
8–7 *Any* [3] *marks from:* Breeding programmes for endangered species [1]; Protection and regeneration of rare habitats [1]; Reintroduction of field margins and hedgerows in agricultural areas where farmers grow only one type of crop [1]; Recycling resources rather than dumping waste in landfill [1].

Trophic levels in an ecosystem (p. 82)

Quick questions

1. A graphical way of representing the relative amount of biomass in each level of a food chain. (Trophic level 1 is at the bottom of the pyramid.)
2. Herbivores that eat plants/algae.
3. Animals that eat herbivores or other carnivores.
4. Carnivores with no predators that are the final organism in a food chain.

Exam-style questions

5–1 Apex predator = tuna [1]; Producer = phytoplankton [1]; Secondary consumer = sardine [1]
5–2 *x*-axis scale symmetrical (up to 400 on each side) and labelled 'Biomass I g m⁻²' [1]; Four bars correct widths (within 1 mm) [2]; Bars correctly labelled with names of organisms [1].
5–3 Loss of biomass = 840 – 12 = 828 so percentage loss = (828 ÷ 840) × 100 [1] = 98.571 % [1] = 99 % to 2 significant figures [1].
5–4 Not all the <u>ingested</u> material is <u>absorbed</u> / some is <u>egested</u> as faeces [1]. Some absorbed material is lost as waste, such as carbon dioxide and water in respiration / water and urea in urine [1].
5–5 Decomposers are microorganisms including bacteria and fungi. They <u>secrete</u> <u>enzymes</u> [1] to digest dead matter and break it down into <u>small, soluble</u> food <u>molecules</u> [1] which can then <u>diffuse</u> into the microorganism [1].
5–6 Eating sardines would mean that there are fewer <u>trophic levels</u> [1] and so <u>less biomass</u> would be lost in faeces/urine/carbon dioxide [1]. Large amounts of glucose are used in respiration, so there would be less <u>respiratory</u> losses [1].
5–7 Reflected by the leaf OR Not absorbed by chlorophyll OR Transmitted through the leaf [1].

Food production (p. 83)

Quick questions

1. Having enough food to feed a population.
2. To increase/maximise growth.
3. Modern farming methods that maximise yields through the use of machines, fertilisers, herbicides and pesticides. (*Accept descriptions of specific techniques, for example, battery chickens, veal pens, caged fish.*)
4. A protein-rich food from fungi.
5. (Human) insulin

Exam-style questions

6–1 *Any* [3] *marks from:* Changing diets in developed countries means scarce food resources from developing countries are transported around the world [1]; New pests and pathogens that affect animals or crops [1]; Environmental changes that affect food production, such as droughts or floods causing crop failure [1]; The increasing cost of agricultural inputs, such as buying machines, pesticides or fertiliser [1]; Conflicts in some parts of the world affect the availability of water or food [1].

6–2 The animals are kept inside in <u>temperature-controlled</u> surroundings [1] and may also be kept in pens or their movement restricted [1]. This means that <u>less energy</u> is <u>transferred</u> to the environment OR <u>less energy</u> is required for movement OR <u>less energy</u> is lost in controlling body temperature [1] so that more energy is available for growth [1].

6–3 *Any* [2] *marks from:*
<u>Unethical</u>/cruel to restrict movement of animals OR to prevent natural behaviour of animals [1]; Diseases could spread more quickly as the animals are close together [1]; Antibiotics are added to food to prevent or treat disease. This can contribute to the development of antibiotic resistant bacteria [1]; Heating the buildings may increase the use of fossil fuels [1].

7–1 Only large fish are caught OR Young fish are small enough to fit through the holes [1] so they can <u>breed</u> and maintain the population [1].

7–2 Introduction of fishing <u>quotas</u> OR Limiting the amount of fish that can be caught OR Reducing the times at which boats can fish OR Limiting the total number of boats [1].

8–1 Fungus/Fusarium [1]

8–2 To <u>kill</u> any microorganisms/ bacteria/fungi that are in the fermenter [1] so they do not <u>compete</u> with Fusarium OR so they do not produce <u>toxins</u> that would contaminate the mycoprotein [1].

8–3 As a source of food OR so the Fusarium can respire [1].

8–4 Provides oxygen for aerobic respiration [1]; Helps the medium to circulate inside the fermenter [1].

8–5 To maintain the optimum temperature for the Fusarium OR because respiration releases heat (energy) [1].

8–6 Fusarium grows <u>faster</u> than animals [1]. Growing Fusarium is <u>more efficient</u> OR uses a small amount of land [1]. Growing Fusarium produces less methane/greenhouse gas than farming cattle [1]. Mycoprotein is <u>healthier</u>/lower fat than meat [1]. Mycoprotein contains fibre unlike meat/eggs/ dairy products [1]. Mycoprotein contains no cholesterol, whereas meat/eggs/dairy products contain cholesterol [1].

Ecology topic review (p. 84)

1–1 *Any* [2] *marks from:*
Temperature [1]; Light intensity [1]; Wind intensity and direction [1]; Availability of water [1]; Mineral content (of sand) [1].

1–2 The waxy cuticle reduces water loss by <u>evaporation/transpiration</u> [1]. The long roots can <u>absorb water</u> from deep in the ground [1].

1–3 The student set up a <u>transect</u> starting from the sea [1] then put down the quadrat every 10 m [1] and counted the number of each plant in the quadrat [1].

1–4 Trend: As the distance from the sea increases, the number of plants per m^2 decreases [1]; Use of data: Only marram grass is found at 10m from the sea / no marram grass at 40 or 50 m [1].

1–5 The marram grass is better <u>adapted</u> to the conditions near the sea [1]. As the distance from the sea increases, the marram grass is <u>outcompeted</u> by rest harrow [1].

2–1 Remaining raised bog = 6% = 7000 km^2; Area of bog lost = 94% = 109 667 km^2 [1]; Rate of loss = 1097 km^2 per year [1]

2–2 Biodiversity is reduced [1] because habitats are destroyed OR food sources are lost [1].

2–3 Greater biodiversity means that the ecosystem is <u>more stable</u> [1] and that there is <u>less dependence</u> of one species on another [1].

2–4 <u>Protect</u> existing peat bogs OR create conditions that <u>regenerate</u> peat bogs [1].

2–5 Can grow crops with: Higher yield OR Herbicide resistance OR Resistance to insect pests OR Drought tolerance [1].

2–6 People may be concerned about the effects of eating GM crops on human health OR The genes might spread into the wild plant gene pool OR; There may be damage to food chains OR there may be a decrease in biodiversity OR Bees/ insects may be harmed by the gene [1].

3–1 Bar chart with 'year' on x-axis and 'area of deforestation' on y-axis [1]; Suitable scales chosen for both axes [1]; Both axes labelled with units for area of deforestation [1]; All seven bars correctly drawn [2].

(*Allow 1 mark for five or six correctly drawn bars*)

3–2 114.2 thousand km^2 or 114 200 km^2 [2] (*Allow 1 mark for 114.2 km^2*)

3–3 *Any* [2] *marks from:*
To provide land for cattle/rice fields/crops [1]; To grow crops for biofuel [1]; To provide wood for building/fuel/paper [1].

3–4 Deforestation can increase the amount of carbon dioxide in the atmosphere through <u>combustion</u>/ burning the wood [1]. Fewer trees means less <u>photosynthesis</u> and so less carbon dioxide is removed from the atmosphere [1]. <u>Decomposition</u> of trees can release carbon dioxide back into the atmosphere when microorganisms <u>respire</u> [1]. Oxygen levels in the atmosphere would decrease because there is less photosynthesis [1]. <u>Methane</u> levels in the atmosphere could increase if the cleared land is used as rice fields or a grazing land for cattle [1]. The amount of water vapour in the atmosphere can decrease because there is less <u>transpiration</u> by trees [1].

3–5 *Any* [2] *marks from:*
Sea ice is melting causing sea levels to rise [1]; Climate change is causing more extreme droughts/ flooding/heat waves [1]; Habitat loss is reducing biodiversity [1]; Migration patterns OR distribution of species is changing [1].

4–1 Control of net size [1]; Introduction of fishing quotas OR reducing the times at which boats can fish OR limiting the total number of boats [1].

4–2 Four symmetrical bars with largest on bottom and getting progressively smaller [1]; Labelled with plankton at bottom and human at top [1].

4–3 Energy/biomass is lost in waste/ faeces/carbon dioxide from the salmon [1]. Energy is transferred to the environment when salmon respire [1].

4–4 The farmed salmon <u>cannot move</u> as much [1] so <u>less energy</u> is <u>transferred</u> to the <u>environment</u> [1] and <u>more energy</u> is available for <u>growth</u> [1]. The farmed salmon are also fed high <u>protein</u> food pellets for rapid growth [1].

4–5 Faeces/uneaten food can cause water pollution OR Pathogens/ parasites from farmed fish could spread to wild fish [1].

Practice exam papers

Paper 1 (p. 88)

1–1 Drawing looks similar to cheek cell and has no shading/sketching [1]; Nucleus and cell membrane both labelled correctly [1].

1–2 Correct magnification = ×1500 (*Accept answers in range 1483 – 1517*) [3]; *If magnification is incorrect: Image diameter correctly measured as 8.9–9.1 cm [1]; Image diameter correctly converted into µm in range 89 000–91 000 [1]; Equation correctly written as image diameter ÷ 60 [1].*

1–3 Cell wall / vacuole [1]

1–4 (site of) protein synthesis [1]

1–5 Prokaryotes / bacteria [1]

1–6 Electron microscope has: Higher magnification [1]; Higher resolution [1]. (*Accept converse answers about light microscopes.*)

2–1 Virus [1]

2–2 The vaccination contains dead/weakened pathogens [1] that stimulate the white blood cells [1] to produce antibodies specific to the pathogen [1]. If the real pathogen enters the body, large numbers of antibody [1] can be produced rapidly [1]. The antibodies bind to the pathogens and destroy them [1].

2–3 *Any* [2] *marks from:*
Small sample size OR only 12 children [1]; No control group OR did not have a group that did not receive MMR [1]; Evidence was based on parents' observations / (what parents said) about their children's behaviour [1].

2–4 6574 or 6575 [1] (*Accept rounding down in this case as you cannot have 0.61 of a child*)

2–5 Concerns about side effects OR Concerns about reports in the media [1].

3–1 *Any* [2] *marks from:*
Used for respiration [1]; Converted into starch for storage [1]; Used to produce fat/oil for storage [1]; Used to produce cellulose for cell walls [1]; Used to produce amino acids for protein synthesis [1].

3–2 Simple sketch graph drawn showing rate of photosynthesis increases as temperature increases, then rate decreases. Temperature on x-axis and both axes labelled [1].

3–3 Number of bubbles of gas (in 10 minutes) [1]

3–4 To check that temperature stays constant / so that temperature does not affect the rate of photosynthesis [1].

3–5 Same type of pondweed OR Same length/size/mass of pondweed OR Same concentration of carbon dioxide in the tube OR Same temperature [1].

3–6 2 [1] (*Answer must be given to a whole number. Do not accept 2.0 or 2.04*)

3–7 x-axis: suitable scale from 0–12 in increments of 2 and labelled 'Light intensity in arbitrary units' [1]; y-axis: suitable scale from 0–60 in increments of 10 and labelled 'Number of bubbles produced in 10 minutes' [1]; All five points correctly plotted [1]; Smooth curve drawn through points [1].

3–8 Answer between 52 and 56. (*Needs to show the idea that the graph is levelling off*). [1]

4–1 Transported from the leaves to the rest of the plant by translocation [1] in the phloem tissue [1].

4–2 Guard cells [1] close the stomata to reduce water loss [1].

4–3 (*Very*) *approximate answer: 1.25 cm³ per hour. Accept between 1.1 and 1.6. Correct answer to 2 s.f.* [3] *marks; Correct answer but to incorrect number of s.f.* [2] *marks; Two values correctly read from graph, but wrong answer for rate* [1] *mark.*

4–4 *Any* [2] *marks from:*
Plant A may be in an environment that has: Higher temperature (than B) [1]; More air movement / higher wind speed (than B) [1]; Higher light intensity (than B) [1]; Lower humidity (than B) [1]. (*Must be comparative statements; Accept converse statements for plant B.*)

4–5 Plants may lose less water if they have: thicker (waxy) cuticle OR fewer stomata OR smaller leaf area OR smaller stomata [1]. *Accept converse statements for more water loss.*

5–1 Coronary arteries [1]

5–2 Less blood can flow through the arteries [1] so less oxygen/glucose is supplied to heart muscle/tissue [1]. Less (aerobic) respiration takes place [1] and less energy is released for muscle contraction [1].

5–3 Serious cardiac event with statins = 6% (6.1149%) [1]; Serious cardiac event with stents = 9% (8.9851%) [1]; (Men aged 60–69 are) less likely to have a serious cardiac event with statins [1].

OR Serious cardiac event with statins is 3% lower than with stents [3].
OR 1.5 times more likely to have serious cardiac event with stent than statin [3].
(*Accept converse statements.*)

5–4 Response should include a judgement, strongly linked and logically supported by a sufficient range of correct reasons. There should be evidence of own knowledge in the answer. For example:
Statins are drugs that reduce blood cholesterol levels and slow down the rate of fatty material deposit in the arteries. This increases blood flow through the coronary arteries and increases the supply of blood to the heart muscle [1]. Stents also increase the blood flow to the heart tissue by inserting a small tube into a coronary artery to keep it open [1]. Taking statins slows the deposit of fatty material in all blood vessels, whereas stents only open a short section of a blood vessel [1].
Both statins and stents reduce the risk of heart attack, but Table 2 shows that statins are more effective than stents at preventing heart attacks and strokes [1]. Statins are taken as tablets and a person will take them every day for the rest of their lives. Stents require surgery, so there is a risk of infection or damage to blood vessels associated with surgery. Blood clots can form around stents, so people with stents would also have to take medication for the rest of their lives to reduce this risk of blood clots. [1].
In conclusion, although there are some serious side effects linked with taking statins, they are more effective than stents. Statins have a beneficial effect on all blood vessels including those in the brain, whereas stents only improve flow to the heart [1].

6–1 Single-celled organisms have a relatively large surface area to volume ratio [1].

6–2 Add Biuret reagent [1] and purple colour shows protein present [1].

6–3 Plasma [1]

6–4 Small intestine is long so more time for diffusion [1]; Villi/microvilli/folds give a large surface area [1]; Good blood supply / lots of capillaries to maintain concentration gradient [1]; Outer surface is one cell thick so short diffusion distance [1]; (Cells have) many mitochondria to release energy for active transport [1].

6-5 Water moves into the lumen/out of the cells by osmosis [1] from a more dilute solution inside the cells OR to a more concentrated solution outside the cell [1] through a partially permeable membrane [1].

7-1 A = bronchus/bronchi [1]; B = trachea [1]

7-2 Two lungs, so 2 × 350 million alveoli; Area of one alveolus = $140\,m^2 \div 700\,000\,000$ [1] = 2 × $10^{-7}\,m^2$ [1]. *Award* [1] *mark for* 4×10^{-7}

7-3 Damaged lung has a reduced surface area for gas exchange [1] so less oxygen diffuses into blood OR less carbon dioxide diffuses out of blood [1].

7-4 The diffusion/concentration gradient is not as steep [1] so less oxygen diffuses into the blood [1]. Less (aerobic) respiration occurs so less energy is available [1] for metabolism/muscle contraction [1].

7-5 Stem cells from human embryos [1] can be made to differentiate into skin cells [1].

8-1 Fatty acids [1]; Glycerol [1]

8-2 Lipids/fats/substrates fit into the active site of lipase/enzyme [1] because the shape of the active site is specific to lipids / because the shape of the active site is complementary to lipids [1].

8-3 Lipids are not emulsified so there is a smaller surface area for lipase to act on [1]. Hydrochloric acid from the stomach is not neutralised so the pH is not the optimum for lipase [1]. The rate of fat breakdown decreases [1]. Fewer fatty acids are available so less fat is stored in the body [1].

9-1 Benign tumours are contained in one area / Malignant tumours invade neighbouring tissues [1]; Benign tumours do not invade other parts of the body / Malignant tumours spread to different parts of the body in the blood [1].

9-2 Age OR Gender OR Stage of cancer OR Fitness [1].

9-3 Water / salt solution / saline containing no drug [1] given through a drip / given directly into the blood stream [1].

9-4 The mouse produces lymphocytes which produce antibodies specific to the injected protein [1]. Lymphocytes are harvested and combined with tumour cells to make hybridoma cells [1]. These are grown in the lab and the ones that produce the correct antibodies are cloned [1].

9-5 With monoclonal antibodies the drug can be delivered directly to cancer cells, so healthy cells are not damaged [1]. However, monoclonal antibodies create more side effects than expected / monoclonal antibodies can cause a severe allergic reaction [1]. Despite the side effects, they can kill cancer cells or stop them from dividing so that the cancer does not spread and form secondary tumours [1].

Paper 2 (p. 98)

1-1 Mayfly nymph [1]

1-2 *Four bars drawn with largest at bottom and getting progressively smaller* [1]; *All bars labelled: pondweed (bottom), mayfly nymph, minnow, kingfisher OR producer (bottom), primary consumer, secondary consumer, tertiary consumer OR trophic level 1 (bottom), level 2, level 3, level 4* [1].

1-3 urine OR faeces OR uneaten parts of the plant/animal OR respiration/as carbon dioxide [1].

1-4 Mean number of nymphs in $100\,cm^3$ = (2 + 3 + 7 + 0 + 8) ÷ 5 = 4 [1]; Mean number of nymphs in $1\,m^3$ $(1\,000\,000\,cm^3)$ = 40 000 [1]

1-5 Volume of pond = $3.14 \times 1.5^2 \times 0.5$ = $3.5325\,m^3$ [1]; Number of nymphs in pond = 40 000 × 3.5325 [1] = 141 300 [1] = 1.4130×10^5 [1]

1-6 Growth of pondweed increases [1] so more food available for mayfly nymphs [1].

2-1 The entire genetic material of an organism [1].

2-2 To search for genes linked to disease OR Understanding and treatment of inherited disorders OR Tracing human migration patterns [1].

2-3 Chromosomes [1]

2-4 X = nucleotide [1]; Y = phosphate [1]

2-5 *All three bases correctly paired* [1]: 1 = C; 2 = G; 3 = T

2-6 DNA is a polymer OR DNA is made up of repeating nucleotides/sub-units [1]; DNA is made of two strands twisted together OR DNA is a double helix [1].

2-7 Could change the sequence/order of amino acids [1] and produce a protein with the wrong shape OR enzyme may no longer bind with substrate [1].

3-1 phototropism OR hydrotropism OR thigmotropism [1].

3-2 Auxin / indoleacetic acid / IAA [1]

3-3 *Diagram shows root growing down* [1] *and shoot growing up* [1]

3-4 Auxin/hormone from the root tip [1] accumulates on the lower side of the root [1] and inhibits/slows cell growth/elongation [1]. More growth on the upper side of the root causes the root to grow downwards [1].

3-5 Roots absorb water from the soil [1]. Water is a raw material for photosynthesis [1]. Photosynthesis produces glucose so more water means more glucose for growth [1]. OR Roots absorb nitrates from the soil [1]. Nitrates are needed for amino acid/protein synthesis [1]. More nitrates mean more protein for plant growth [1]. OR Roots absorb magnesium from the soil [1]. Magnesium is needed for synthesising chlorophyll [1]. More chlorophyll means more light can be absorbed for photosynthesis [1].

4-1 The ciliary muscles relax [1] increasing their diameter [1] and pulling the suspensory ligaments tight [1]. The lens is pulled thin [1] and only slightly refracts the light rays [1] so they meet and focus on the retina [1].

4-2 The eyeball is elongated/too long OR the lens is too thick/curved [1] so light is focused in front of the retina [1].

4-3 Impulse travels along a sensory neurone to the brain [1] then along a motor neurone [1] to muscles in the eyelid [1].

5-1 Long wings [1]

5-2 Dd [1]

5-3 *Parental genotypes correct = Dd and Dd* [1]; *Offspring genotypes correct in Punnett square = DD, Dd, Dd and dd* [1]; *Offspring phenotypes correctly linked to genotypes = DD and Dd long wing, dd short wing* [1].

5-4 25% [1]

5-5 (nucleus removed) so the egg cell only contains DNA from the cell to be cloned OR the egg cell does not contain DNA from the egg donor [1]; (electric shock) to stimulate the egg cell to divide [1]; (sterile) to prevent growth of microorganisms OR to prevent disease [1]; (culture medium) to provide glucose for respiration OR to provide amino acids for protein synthesis [1]; (25 °C) to provide the optimum temperature for growth/enzymes [1].

5-6 Genes/DNA/nucleus from fly with short wings [1].

5–7 Fruit fly and spotted wing drosophila are the <u>most closely</u> related [1] because they are the <u>same genus</u>/family [1]. OR Fungus gnat is the <u>most distantly</u> related [1] because it is a <u>different family</u> [1].

6–1 ADH [1]

6–2 Pituitary (gland) [1]

6–3 <u>ADH</u> is released by the <u>pituitary</u> gland when the blood becomes too concentrated [1]. ADH increases the <u>permeability</u> of the kidney tubules [1] so that more water is <u>reabsorbed</u> back into the blood [1].

6–4 *Any [4] marks from the following; must contain a judgment, strongly linked and supported by a range of correct reasons including pros and cons.*
Dialysis requires repeated treatments whereas a transplant is a one-off procedure [1]; People on dialysis have a restricted food and drink intake, but people can eat a normal, healthy diet after a transplant [1]; Dialysis requires needles to be inserted in the skin so there is a risk of infection every time dialysis happens [1]; Urea and other toxins can build up in the body between dialysis sessions [1]; There are serious risks to having transplant surgery including blood clots or infection [1]; Taking immunosuppressant medication after a transplant means you are more at risk of infection for the rest of your life [1]; There is a long wait for a transplanted organ and often someone has to die for an organ to become available [1].
Plus conclusion for final [1] mark:
In conclusion, despite the risks associated with transplant surgery and having to take immunosuppressant medication, it is worth having a kidney transplant to be able to live a normal life without the dietary restrictions and risk of infection involved with dialysis.

7–1 More bacterial cells so more (aerobic) respiration. [1]

7–2 *Tangent drawn to curve* [1]; Rate at 10 hours = 9.5 [1]; Rate at 16 hours = approximately 3.5 [1]; (9.5 ÷ 3.5) = approximately 2.7 times faster at 10 hours [1].

7–3 *Any [3] marks from:*
Glucose/sugar used up so no respiration to provide energy for growth [1]; Amino acids/protein used up so no more growth [1]; Toxins produced by bacteria build up [1]; Respiration increases temperature above optimum [1].

8–1 Use a $1\,m^2$ quadrat and place it <u>randomly</u> in the field using a random number generator to generate coordinates [1]. Count the number of maize plants in the quadrat, then <u>repeat</u> at least nine more times [1]. Find the <u>mean</u> number of plants per quadrat (per metre squared) [1]. Find the area of the field, then multiply this by the mean number of maize plants per m^2 to find the size of the whole population [1].

8–2 Glufosinate sodium can be used to kill weeds, but will not kill the maize [1]. The maize does not have to <u>compete</u> with the weeds for light/minerals/water [1], so the crop will grow better and have a higher yield [1].

9–1 FSH from the pituitary gland causes maturation of an egg in the ovary [1]. Oestrogen from the ovary causes thickening of the uterus lining [1]. Oestrogen inhibits the release of FSH and stimulates the release of LH from the pituitary gland [1]. LH stimulates the release of an egg from the ovary [1]. Progesterone from the ovary maintains the uterus lining and inhibits the production of FSH and LH [1].

9–2 Body temperature rises sharply OR rises from ~36 to 37 °C on day 14 [1] which is when the level of LH is the highest so ovulation occurs on day 14 [1] and there is more likely to be an egg there to be fertilised [1].

9–3 *Any [3] marks from:*
Age [1]; BMI/body mass [1]; How long they had been trying to conceive [1]; Number of previous pregnancies/children [1]; History of smoking [1]

9–4 To prevent false claims OR To check the quality of the results OR To check the conclusions are valid OR To establish a consensus about which claims should be regarded as valid [1].

9–5 *Any [5] marks from the following:*
Higher number of pregnancies with artificial insemination than clomifene citrate/no treatment so headline is not supported [1]. The p values were above 0.05 so no significant difference in the number of live births with either treatment which supports the headline [1]. There were more side effects with clomifene citrate than artificial insemination/no treatment which supports the headline [1]. There was a large group of women in the study and there were similar numbers of women in each group [1]. The study used a control group for comparison [1]. The study was peer reviewed before it was published so other scientists have checked that the results and conclusions are valid [1].